# 4 Steps to Solving Your Problem

The ONLY Troubleshooting Resource

You Will Ever Need

*By Chris Abbott*

# 4 Steps to Solving Your Problem
© 2017 Chris Abbott

All rights reserved. No part of this book shall be reproduced, stored in a retrieval system, or transmitted by any means, electronic, mechanical, photocopying, recording, or otherwise, without written permission from the author. No patent liability is assumed with respect to the use of the information contained herein. Although every precaution has been taken in the preparation of this book, the publisher and author assume no responsibility for errors or omissions. Nor is any liability assumed for damages resulting from the user of the information contained herein.

ISBN-13: 978-0-9996643-0-8 (Paperback)
ISBN-13: 978-0-9996643-1-5 (EPUB)
ISBN-13: 978-0-9996643-2-2 (Mobi)
ISBN-13: 978-0-9996643-3-9 (PDF)

First Printing: December 2017

Warranty
Every effort has been made to make this book as complete and as accurate as possible, but no warranty or fitness is implied. The information provided is on an "as is" basis. The author and the publisher shall have neither liability nor responsibility to any person or entity with respect to any loss or damages arising from the information contained in this book.

Editor: Janette Jones
Proofreaders: Jodie Smith, Judy Slaughter, Sam Harding
Cover Designer: Jill Jasuta, jilljasuta@gmail.com


Media Inquiries, speaking engagements or to contact the author directly:
email: chrisabbottauthor@gmail.com
Phone: 443-235-7290

# TABLE OF CONTENTS

About the Author .......................................................................... iii
Introduction ................................................................................... v
**Chapter 1 We've All Been There** ............................................... 1
    Learn, Identify, Isolate, and Conquer ......................................... 3
    Do It Now .................................................................................... 6
**Chapter 2 Fundamental Relationships** ..................................... 9
    Music Always Helps .................................................................. 11
**Chapter 3 Learn, Learn, Learn** ............................................... 17
    Working Knowledge ................................................................. 18
    Input, Processing, and Output .................................................. 20
    Systems Made of Systems ........................................................ 24
    Environmental Factors .............................................................. 27
    How is it built? .......................................................................... 29
    Installation and Configuration .................................................. 31
**Chapter 4 Resolution Skills** ..................................................... 33
**Chapter 5 People** ....................................................................... 41
    The Genuine Helper .................................................................. 44
    The Honestly Ignorant ............................................................... 45
    The Helpee Helper .................................................................... 46
    The Know-It-All ....................................................................... 47
    The Expert ................................................................................. 48
    The Blame Seeker ..................................................................... 49
    The Dribbler .............................................................................. 50
**Chapter 6 Identify the Situation** .............................................. 53

    Four Out of Five Senses Agree .................................................... 54
**Chapter 7 Isolate the Problem ................................................ 63**
    Questioning Change .................................................................. 67
    Answer Me This ........................................................................ 70
    But what if...? ............................................................................ 72
**Chapter 8 Conquer .................................................................. 77**
    Unintended Consequences......................................................... 81
    Inside-Out.................................................................................. 82
    Always Think, But Not Too Much ............................................ 84
**Chapter 9 War Stories.............................................................. 85**
    Cloudy Communications ........................................................... 86
    Aaaaand We're Down ............................................................... 89
    Sometimes You Have to "Vent"................................................ 93
**Chapter 10 Tips and Advice .................................................... 99**
    Customers.................................................................................. 99
    Tools........................................................................................ 104
    Step by Step............................................................................. 107
    Spotting Trends ....................................................................... 108
    Observe................................................................................... 110
    Documentation ........................................................................ 110
    Working Well With Others...................................................... 111
    Rules, Not Laws ..................................................................... 112
    You Can't Win Them All ........................................................ 113
**Chapter 11 In Conclusion ...................................................... 115**

# ABOUT THE AUTHOR

CHRIS ABBOTT CURRENTLY serves as a Computer Network Specialist for the State of Maryland's Department of Health and has over 25 years of experience in the information technology industry. He has extensive knowledge in systems design and administration, consulting, customer support and training. Chris has a wide variety of interests including music performance and production, sports, motorcycles and fishing. When not working or pursuing his interests, Chris can be found at home with his family in Dorchester County on Maryland's beautiful Eastern Shore.

# INTRODUCTION

IF THE ROLE of tech support representative, service provider, or technician is your chosen career path, then this book is definitely for you. Let's face it: problem solvers, tech support specialists, service providers, and anyone who works in a career focused on end user support and repair are the unsung heroes of modern society. Without them, simple problems can become huge undertakings to navigate and resolve, and, even worse, catastrophic failures can result in unnecessary downtime and lost revenues for your client.

Have you ever noticed that every service manual or user guide you read always has a troubleshooting section in the back? It usually contains a neat little table that lists specific problems as well as some potential solutions for the most basic issues you may have with a given product. Even most, if not all, do-it-yourself books spend most of their pages providing you with step by step instructions on how to repair specific items. They never seem to give you the knowledge needed to properly troubleshoot and repair a problem in general. That's where this book is different.

*Introduction*

Throughout this book you will discover the one common method that all successful service providers use, knowingly or unknowingly, to resolve any and all problems they face. Best of all, this method of problem resolution is not industry specific. That means you can use the same techniques no matter if you service computers, appliances, heating and air conditioning systems, automobiles, or anything in between.

This method, known as *Learn, Identify, Isolate, and Conquer*, isn't some fancy new shortcut to career immortality but rather a clear and concise method to help you develop your troubleshooting and problem resolution skills. The knowledge and material you will be exposed to in this book is not based on theory or conjecture, but instead is compiled from my over 25 years of experience as a service provider in a number of different disciplines as well as information gleaned from colleagues in various other industries. Even if you are a seasoned professional service provider with years of experience in your field of expertise you will find the material in this book valuable and insightful.

But what if you are not currently a service provider but you are interested in that career path? Or perhaps you never intended to be a service provider but were somehow thrust into the role as part of your current occupation. Or maybe you are just an average, everyday person who likes the idea of being able to fix things yourself. After all, it can be fun and rewarding while at the same time allowing you to save money on repair bills. If you fit into any of these categories then this

*4 Steps to Solving Your Problem*

book will definitely lead you down the path to troubleshooting and problem-solving greatness!

The most important thing you need to do is LEARN. The chapter on Learning is without a doubt the most vital chapter in the book. The primary focus is for you to build what is known as a "working knowledge" about, well, everything. It sounds difficult but it is much easier than one might imagine. You will learn that in order to gain this working knowledge you will need to figure out the three basic steps of operation that all things have in common. These steps are known as Input, Processing, and Output. It is extremely important to grasp this concept because, as you will see, the operation of everything you come into contact with on a daily basis can be broken down into these three fundamentals.

Another concept you will need a firm understanding of is once you identify how a device works you will also need to understand how to identify the different individual systems that perform each part of that process. In addition, we will discuss how seemingly different devices share numerous similarities in both function and operation even though they are performing vastly different tasks.

If you are a service provider or it is your intent to become one, you will need to learn a few people skills before we go any further. No matter how shiny or effervescent a personality you possess it will do you no good unless you are able to relate to the needs of your customers. And trust me on this, there will be needs! You will need to learn how to interact with your clients

*Introduction*

and you will need to be able to decipher their various ways of trying to explain what it is that they are in need of. More specifically, you will need to learn to be at ease around some of the more interesting types of people and their associated personalities you will encounter in your daily routine.

Now, don't go getting all discouraged on me! You will meet some great people throughout your career. Conversely, you will meet some not so great people as well. Chapter 4 deals with a variety of characters you will run into throughout your travels, with advice on how to cope with their individual peculiarities. In doing so you will quickly discover that not only will you function as a service provider but also a psychologist, a social worker, and, at times, even a babysitter.

Next you will learn how to properly identify what is actually occurring. Understanding the exact nature of the problem is crucial to the rest of the process. Much care should be spent properly identifying what is happening (or not happening) prior to attempting to isolate a cause. You will also learn how to identify similarities between seemingly dissimilar devices as well as how to apply knowledge of one device to another.

You will then need to isolate the root cause of the problem. To do this you will implement a technique of eliminating possible causes based on your working knowledge of the device as well as the information you obtained while identifying the actual problem. It is at this moment when your working knowledge of the individual systems used to perform the IPO tasks identified during your learning process is most important.

And finally, you will conquer the problem. You will not only repair the damaged device but you will learn to examine the environment in which it resides to determine the nature and/or severity of the needed repair. Remember, duct tape can only do so much!

Great effort has been taken to explain the material using plain English—with a little humor thrown in—while avoiding technical jargon where possible to make sure that you, the reader, can easily understand the process as well as the various methods used to implement it. I have also included many real-world examples in each section to give you a better idea of how I implement this method myself. My hope is that you will find the information in this book informative and extremely useful and that you will continue to use it as a reference guide as you flourish in your career.

Cheers!

# CHAPTER 1
# WE'VE ALL BEEN THERE

IT'S 35 DEGREES outside, it's raining, and you are 20 minutes away from somewhere you needed to be 10 minutes ago. You get in your car, put the key in the ignition, turn it, and nothing happens. You try a few more times but to no avail. You begin to get fidgety and nervous. Within seconds panic starts to set in. You check the lights, the radio—all works fine. You pop the hood and get out of the car into the now freezing rain and begin to look for anything out of the ordinary even though you have no idea what ordinary actually looks like. You jiggle a few wires, make sure the battery is connected, and then get back into your car, cold, rain-soaked, and near the edge of a nervous breakdown. You make that dreaded call for help and then sit and wait anxiously for help to arrive. Not only are you now late for your appointment but you are cold, drenched, and about to receive a bill that could be hundreds of dollars, or worse!

Help arrives in the form of a big truck with flashing amber lights that might as well be emitting dollar signs instead of pretty amber beams. You explain to the tow truck driver that it just won't start. He gets in your car, looks around for a second, puts your car in Park, turns the key, and miraculously brings your car back to life. All for the low, low price of $75!

"Looks like you had 'er in Reverse," he says with a grin. He then informs you that cars will not start unless in Park or Neutral and some won't even start in Neutral. "If all you did was put it in Park why are you charging me $75?" you ask. "Well, I put it in Park and started it for free. The $75 is for me knowing what to look for," he replies. "Have a good night," he says. If only you had that information before this situation and not after!

This story, as fictional as it may seem, is all too real for some people. Most problems we encounter in our daily lives, be they at work or at home, can be resolved very easily if we equip ourselves with the appropriate knowledge ahead of time. For example, if you had learned a few basic things about how a car actually operates, one of the first things you would have checked in the situation above was to see if the car was in Park. Once you noticed the gear shift was in Reverse you would have moved it to the Park position and then been able to start the car. This is but one example of how my method of troubleshooting and problem resolution can help you in your career as well as at home. With that in mind let's take a look at how we can *"Learn, Identify, Isolate, and Conquer."*

## Learn, Identify, Isolate, and Conquer

How many times has each of us been in a situation where we had no idea what was going on, no idea what we were dealing with, and no clue how to resolve it? Even worse, when asked what was happening by someone who was charged with resolving the situation, we did not even possess the ability to explain the problem accurately. Instead we rambled on incoherently and provided the person with a barrage of nonsense that Einstein himself would need a translator to figure out.

You say to yourself, "Well, I can't know everything! It would be impossible to know how to fix everything or know how everything works." Believe it or not, this statement is both true and false.

You would be correct to say there is no way you could know every little detail about everything. In fact, in my experience, even the experts who specialize in a particular discipline or occupation do not know everything about the area of knowledge in which they specialize. What is not impossible for you is to LEARN how to obtain the knowledge that would allow you to enter a completely foreign situation, IDENTIFY or assess the situation, ISOLATE the root cause, and, if not CONQUER it yourself, be able to communicate effectively to the person who can.

How is this possible, you ask? It's easier than you might think, but like anything worthwhile, it takes time and effort and a bit of an open mind.

*We've All Been There*

Throughout my career I have learned that the one skill set everyone needs in order to be successful in their occupation—or life in general—is the ability to properly identify a problem and take the necessary steps to resolve it. I know it sounds obvious, and should be common sense, but you would be extremely surprised at just how incapable most people are at this very task. Instead of thinking clearly and analyzing a situation properly they enter panic mode and, in the process, their thinking becomes scattered and disconnected from what is happening. Not only is this completely counterproductive but, depending on the situation, potentially dangerous.

The method I share in this book has helped me resolve just about every situation I have ever encountered in my life and career. Whether it's the most mundane issues you face in your day to day life or the most stressful, employing this method will help you to overcome these problems and everything in between calmly and efficiently. In fact, over the years, I have used this method in so many situations that I have been accused of having a rather nonchalant attitude toward the very crisis I was in the middle of resolving. In actuality, using this method has allowed me to learn how to keep my stress levels to a minimum. I have trained myself to focus my energy on identifying and resolving the issue at hand and not on what might happen next should I not be able to solve it. You can train yourself to do this as well. It just takes a bit of time and patience as well as a dash of confidence.

*4 Steps to Solving Your Problem*

Another very useful thing you will learn is the ability to identify fundamental relationships. Understanding fundamental relationships is the key to being able to use knowledge of one particular device to assist you with using another. This simply means that once you learn how a particular device works you can then use that knowledge to figure out how similar items function. This idea of fundamental relationships can be easily explained by looking at the cut-copy-paste functions on a computer. We all know that the cut-copy-paste commands work on every program and function of a computer and that these functions cannot only be used to accomplish a multitude of tasks but can also be accessed in a myriad of ways. We can access these commands through menus or by way of keyboard shortcuts in any combination we choose. What some of us may not know is that even if we do not see an option to use these commands, as long as we know all the different ways we can access them, in most cases, we can still use them successfully.

Fundamental relationships can be found between different career paths as well. The idea that you can apply your existing knowledge and/or set of skills to a new career in a completely different area of expertise is very difficult for some people to grasp. The inability to identify these relationships is effectively the biggest stumbling block most people face. Let's say you have a degree in education and are currently employed as a schoolteacher. Did you know that you can use the same skills you use as a schoolteacher to be a corporate trainer, an event planner, or even a business manager? The knowledge, management, and organizational skills you use to plan your lessons, present the material to your students, and work with them to complete their tasks are the same skills needed to put on a wedding or other event or manage a staff of employees to accomplish your goals in business. The only difference is in the manner in which these skills are applied.

**Do It Now**

The late Benny Hill, a famous British comedian from the 1970s, while reciting one of his infamous limericks, said, "Why put off until tomorrow what you can do today, for it may lead to great pain and sorrow, because if you do it today and like it, you can do it again tomorrow!" I'm pretty sure he was not referencing projects or tasks for your job or even household chores, but the premise is clear. If you learn nothing else from this book you need to learn this. Do it now!

A wiser man than me once stated that although it's never too late to start learning, why would you want to put it off? If there is one thing I cannot understand it's why people procrastinate, sometimes to the point of not starting at all. They have every reason why they can just do it later. Do it now!

People will say, "I always work better under pressure," or "I work better when I'm near a deadline." I call BS on this! These are just excuses for being lazy, period. I know that because I used to be of the exact same mentality. When presented with a deadline I figured I had all the time in the world to accomplish the task. I would work harder at avoiding what needed to be done than if I had just sat down and done the work. Invariably, as the deadline neared, I would dive in head first and frantically try to get everything done at the last minute, only to find that some of the necessary processes were, in reality, going to take much longer than I had anticipated. Needless to say, I missed my deadline. It wasn't long before I learned the old adage of "work first, play later" and just how important those four simple words are. There is no excuse for not starting but I can assure you there are consequences. Don't put off learning until later! Do it now!

I challenge you to put into place the methods you will learn from this book as you are reading it and not afterward. Not only will doing this be beneficial to your life but it also makes the process of learning and understanding the steps in this method much easier. If you are like me you learn more about something by actually getting your hands dirty than you do by reading about it. If you follow the steps I have outlined for you it is my personal belief that your professional and personal lives will benefit greatly.

You must simply do it now!

# CHAPTER 2
# FUNDAMENTAL RELATIONSHIPS

THE QUESTION I have been asked the most throughout my career is, "How are you able to walk into a situation where you have no experience with what they have, how it's set up, or how they use it, and resolve whatever issues they are encountering in a timely manner?" To properly answer that question, it would be good to know a little about me.

For most of my adult life, as well as in my early years, I have been a problem solver. In my professional career, I am an IT consultant with over 25 years of experience in the field. In addition to that I have also worked as a live sound and studio engineer (for a number of local bands in my area), a musician, and an auto mechanic, and I've done numerous other little side jobs helping others in need. I've done stints in fast food, retail, and manufacturing, and have even spun wrenches on airplanes and race cars. I am also hopelessly addicted to a lot of the "How-To" TV shows and resources found on the internet.

For me, it was never enough to know that a particular device could perform a certain function. I needed to

*Fundamental Relationships*

know HOW it did it. I have always been fascinated by how things work as well as the process in which things are manufactured. When I was a kid I could always be found deconstructing my toys and attempting to put them back together. In high school I enrolled in our local vocational-technical school and studied basic electronics. I also spent a lot of time hanging out at one of the local auto repair shops and at a local boat dealer's service department just to experience and learn new things. It's a behavior I find myself doing to this day.

Throughout my life I have understood two simple rules. Rule number one is you will never know everything. Rule number two is you will never know enough. It wasn't until later in life that I discovered a third and, arguably, more important rule. That rule is that fundamental relationships are the building blocks of life. Understanding this is the key to a successful career as well as a happy and prosperous life.

So, what is a fundamental relationship? Well, to properly answer that we need to know what the word fundamental means. As an adjective, the word fundamental means being an essential part of, a basis or an original or primary source. As a noun, it is defined as a basic principle, rule, or law that serves as the groundwork of something. A fundamental relationship, in this sense, can be described as a number of different disciplines that share a common foundation. It's an idea that many different things, be they theories, products, or even entire industries, share a common framework and source. In order to be a trusted and successful service provider you must be able to comprehend the most basic principles, logic, and functionality associated with the discipline in which you specialize. Once you are fluent in the basics you will have the ability to recognize commonalities shared between similar and dissimilar things.

**Music Always Helps**

The best way to describe the concept of a fundamental relationship is with a quick music lesson. If we break music down to its core elements we see that there are 12 individual music notes (A-A♯-B-C-C♯-D-D♯-E-F-F♯-G-G♯) and that they are arranged in various groupings, usually consisting of seven notes, called scales. For you musical purists out there, I purposely omitted the associated flats ( ♭ ) for these notes in the interest of simplicity. Please forgive me.

Our lesson today begins with constructing a simple major scale. To do this we need to understand a few things. First, if we look at our list of musical notes the distance between each note is called a half step and the distance between two notes is a whole step. So if we look at the note A, the next note in the sequence is A♯ (the "♯" symbol in musical terms is referred to as a "sharp"). This is considered a half step interval. If we look at the note A and then go to the note B, this is considered a whole step interval. All musical scales are constructed using a predetermined series of half and whole step intervals. For example, our major scale would be constructed with the following steps.

*Fundamental Relationships*

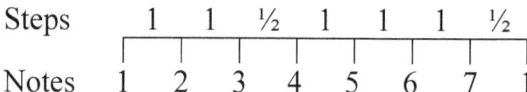

Notice that there is a whole step interval between the first and second notes as well as the second and third notes, but then only a half step interval between the third and fourth notes. Now let's substitute the numbers in the notes line on the bottom of our diagram and see what notes we would use to construct a C major scale.

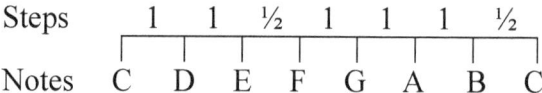

Referring back to our list of notes and starting with C, if we move a whole step we jump over the C# note and land on D. If we move another whole step we land on E. Notice the next interval between notes is only a half step. Looking at our list a half step from the note E is the F. Now let's construct an A major scale using the same procedure.

*4 Steps to Solving Your Problem*

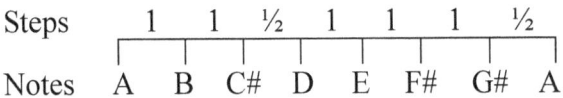

Okay, great! Now why is this important? Well, I'm using this little music lesson to show you how learning one simple thing, like a major scale in music, can actually lead to a greater understanding of a multitude of things. For example, notice how we created an A major scale in the same manner in which we created the C major scale. We used the same pattern of whole and half steps but we simply started on a different note. So, our fundamental relationship lesson for today is that by learning the 12 notes of music and how to construct a major scale using a pattern of whole and half steps, we can apply this knowledge to learning 12 major scales! All we need to do is pick a starting note and then apply the major scale pattern to find the remaining notes in the scale.

Let's take our musical example a step further. Let's say you are a musician and you play the guitar. You know all of your major and minor scales and can pretty much play any song you are presented with without too many problems. Now someone hands you a bass guitar to play. Your guitar has six strings but this bass only has four. Can you play it? The answer is, of course you can. You know there are only 12 notes in the musical vocabulary and you know how to build scales from those notes. And guess what? The bass guitar, which

*Fundamental Relationships*

only has four strings, is tuned to the exact same notes as the lower four strings of your guitar. That means you can directly apply your musical knowledge of the six-string guitar to the four-string bass with no effort at all.

Let's consider another example of fundamental relationships at work. Imagine that you are an auto mechanic. You understand how a gas combustion engine converts fuel into energy. You understand how all the systems that assist the engine operate and you know how all the systems unrelated to the engine operate. Best of all you can successfully troubleshoot and repair any problem presented to you in a timely manner. Now let's say someone brings you a boat with an engine problem. The engine runs but not very well. You have never worked on a boat engine before. Using your abilities as an auto mechanic, do you think you would be able to troubleshoot the boat's engine and resolve its problem? If you answered yes, then you are, indeed, correct.

I know what you are asking me. How can an auto mechanic who has never worked on a boat engine before know how to troubleshoot one? The answer is quite simple. It's because a car engine and a boat engine function in the same manner. Sure, there are some differences between the two, but their method of operation is identical. In this case the knowledge of how to troubleshoot and repair a car engine can also be used to find the problem with the boat's engine because they share a fundamental relationship.

The secret to being a successful and skillful service provider is to know the fundamental relationships that

*4 Steps to Solving Your Problem*

all things have in common. For example, if your job requires you to install or repair electrical devices of any kind, be they computers, appliances, or simple light switches, you would benefit from taking a course or getting a self-study guide on basic electronics. Why? Because basic electronic theory and concepts are the building blocks for all things electrical, that's why. Knowing this information will help you to understand concepts like signal flow, resistance, and polarity, as well as many other topics. Knowing that there is a fundamental relationship between basic electronics and all electrical devices ever created will not only make it easier for you to understand new and emerging technologies, but it will also help to develop you into a leader in your field.

As I stated earlier, fundamental relationships are the key building blocks which many related—and sometimes seemingly unrelated—theories, products, and industries are constructed upon. Think about this for a second. If you removed those 12 musical notes we discussed earlier in this chapter from society the entire music industry that depends on using those 12 notes in a creative (or not so creative) fashion would no longer exist. As you read the rest of this book try to recognize the fundamental relationships at work in each of the examples that are presented to you. Doing this will help you to understand that everything we use in our daily lives, no matter how complex, has a fundamental relationship with many other things. Understanding that relationship, be it ever so complex and diverse or remarkable in its simplicity, is the key to being successful in your career as a service provider.

# CHAPTER 3
# LEARN, LEARN, LEARN

IN THIS DAY and age, it cannot be stressed enough that the simple act of learning is key to our daily survival. In fact, I truly believe that if there is ever a day when I do not learn something, that day can be considered a complete and total waste. If you think about it we constantly learn new things every day whether we intend to or not. We learn what is going on in the world by reading or watching the news. We learn what our friends are doing (or not doing) by browsing through the various social networks on the internet. Most of what we learn we immediately forget, partly because we consider it useless information or simply because we do not spend much time with it. If you are like me you have more important stuff to do and need to move on to the next thing in your day. In the end, any way you look at it, the one thing we must do in life is learn.

Step one of our learning process couldn't be simpler. You just need to learn everything you can about everything! That sounds impossible, doesn't it? Well, it's actually easier than you might think. Understand that we do not need to know every little detail about every little thing we come into contact with. We are more interested in what is known as a working knowledge of things rather than a detailed knowledge of them.

**Working Knowledge**

A working knowledge of something can be defined as an overall understanding of how similar things work. A good example of this would be to know how a car operates. Most of us simply know how to drive a car, how to put gas in it, and how to take it to a mechanic to get the oil changed once in a while. However, when trouble arises we are clueless as to how the car actually operates and therefore unable to accurately troubleshoot the issue or coherently describe to a mechanic what we are experiencing.

You don't need to know all of the minute details about a car to have a good working knowledge of how one operates. And once you learn how your car operates you have also learned how most cars operate because they all share the same fundamental relationships. Simple.

So just how are most cars similar? All cars that use a gas combustion engine or a diesel engine function in much the same way. They all share similar components like a steering wheel, brake and accelerator pedals, climate control, windshield wipers, etc. They also have an engine that burns fuel and uses oil for lubrication, and water or antifreeze that circulates through a radiator to keep it running cool. The similarities don't end there. They all have a fuel tank with a line that carries fuel to

*4 Steps to Solving Your Problem*

the engine compartment. Fuel injectors pass a mixture of fuel and air into the cylinders where it is compressed and ignited, and then the exhaust is ejected into a manifold and carried away in a set of pipes, through a muffler and out the tailpipe.

Once you learn how to drive one car you are then capable of driving most other cars as well! You get the idea.

Now that you see the basic similarities that all cars share I would like you to use that same principle to look for similarities in other things. Take some time to investigate and compare the following devices.

- Refrigerator and air conditioner
- Hair dryer and clothes dryer
- Dishwasher and clothes washer

Pay particular attention to their similarities, identify their fundamental relationships, and also note their differences. You will notice that, for the most part, their differences are merely modifications to an identical process in order to accomplish a specific task. Understanding this information is the key to building a functional working knowledge as well as developing our shared skill sets.

*Learn, Learn, Learn*

## Input, Processing, and Output

Now that we know how to define what working knowledge is, just how do we go about acquiring it? Would you believe that there is a simple process you can use to help you figure out how everything operates? This approach, called IPO or Input-Processing-Output, is used in the technology industry for systems analysis and software engineering. It is used to describe the structure of how a piece of software or a process works. In reality, this approach is the key to how everything on planet earth operates. As simple as it may seem, it's also one of the most highly misunderstood.

This is, without a doubt, the most basic principle you need to understand and at the same time the most difficult. I say that because once you use this approach to understand the basics of how one thing works you will then have the ability to apply that knowledge to a myriad of other things. The simple fact is everything in life depends on these three operations, and it is important to remember that these three operations do not change.

For anything to function there needs to be a method of input. This can be an action we perform to initiate a task such as flipping a switch to turn on the lights. Or it can be something that happens automatically in response to certain criteria such as a motion sensor activating the lights when it sees movement. We need to figure out what constitutes the input medium for the device to operate.

*4 Steps to Solving Your Problem*

Next, we need to figure out how that input is processed. What happens to the input medium during the process of operation and how is that operation carried out? In this case flipping the light switch closes an electrical circuit and provides power to the light bulb causing it to emit light. The motion sensor performs the same operation but does it automatically when movement is detected.

And lastly, what is the desired output? If we flip that light switch then the desired result is light to illuminate the room. If we have a motion sensor on a light we can assume that our intention is to automatically illuminate the area when movement is present.

Let's look at some other examples. How does a computer operate? A computer takes input from a keyboard, mouse, or other device, processes that information using different types of software, and then outputs the results to a screen, a printer, or a file stored on your hard drive.

Here is an example on a somewhat grander scale. Have you ever been to a large event or a concert and wondered how the sound or public-address system works? A PA system uses microphones or instruments as input devices, the mixing board handles all of the processing and routing of the audio signals, and finally those audio signals are output to a series of amplifiers and speakers.

Earlier we learned about the similar components that all cars share. But how does a car operate? Very simply stated, a conventional car uses gasoline as an input medium that is processed by the engine and converted into power with exhaust being the output.

Even we humans follow this rule in a number of different ways. We input information to our brains by means of what we see, hear, smell, touch, and taste, and our brains process the information with the output being some sort of action. Another example would be the fact that we eat food, our body processes that food for energy, and the output usually requires a trip to the bathroom.

So how do we go about actually identifying how a device operates using the IPO approach? In order to identify the Input, Processing, and Output operations of a device you need to answer a few basic questions. How do I activate the device and what does the device need in order to operate? What happens when I activate the device? What is the outcome or result? While you are finding the answers to these questions watch and listen intently to what the device is doing and the noises it generates. This will all be very useful information should you need to troubleshoot a problem with it later.

Let's take a look at a rather simple example. Let's take a look at how a toilet works. We all know how to operate a toilet but we may not know how the toilet itself operates. A simple observation of our toilet reveals a bowl filled with water, a seat with a lid, and a tank with a handle. If we want to know how it operates we need to be able to observe what happens when we push the handle down to activate it. Since the handle is located on the side of the tank there is a good chance that inside the tank is where we will find what we are looking for. We remove the tank lid and gaze into the watery abyss. Now what?

Well, next we need to activate the toilet by pushing the handle and then paying close attention to what is taking place inside the tank and the bowl. That is our Input. (Technically speaking our input would be to… well, never mind.) Pushing the handle to start the cycle of flushing the toilet is step one. Next, we need to figure out how the toilet processes the input. (In the spirit of this book's intent, and its PG rating, I will refrain from any more toilet humor.)

What happened when you flushed the toilet? If you were watching closely you would have noticed that the stopper at the bottom of the tank is attached to the handle via a chain or some other type of connector. Pressing the handle lifts the stopper, which is attached to the tank via a hinge, thereby releasing the water stored in the tank into the bowl below. When the tank is empty the stopper falls back down over the drain hole and water is pumped into the tank by way of the filler valve. As the water level rises, a float attached to the filler valve rises with it, and when it reaches a specified level it closes the valve and the tank is now full. This is our Process. The Output would be the removal of the contents of the bowl into the sewer pipe, leaving us with a bowl full of clean water. That wasn't too difficult, now was it?

By using the IPO approach, we can begin to build a vast working knowledge about everything around us, and as we do we will begin to see similarities in how completely different devices function as well as their fundamental relationships. More on that later!

**Systems Made of Systems**

So, let's break things down a bit further, shall we? Just what is a "system" anyway? Well, a system is a set of connected individual items that creates a complex device or process. Nearly every device you come into contact with is actually a group of specialized systems working together as a unit to complete a particular task.

I'm sensing you need an example here.

Have you ever watched a football game? Let's say a football team is a device. That device is comprised of a number of specialized systems called players. Each system, or player, is assigned a specific task that it must do in order to accomplish a particular goal. If one of those systems malfunctions or, in this case, doesn't perform the task correctly, then the device breaks down and does not accomplish its mission.

So how do we determine what each system does? That's where the IPO method takes over. We can use this method to determine how the device operates as a whole, as we have seen in earlier examples, as well as to determine how each individual system operates and what function each individual system contributes. For instance, the system, or player, known as the center is responsible for putting the ball in play by handing it to the system known as the quarterback. So, the intended IPO process for the center would be as follows:

***Input:*** The center takes his position awaiting input from the quarterback. In this case the input would be in the form of a word spoken (or more appropriately, yelled) by the quarterback informing the team to initiate play.

***Processing:*** Upon receiving the required input the center snaps, or hands off, the ball to the quarterback and begins the task of defending him from oncoming defensive players from the opposing team.

***Output:*** The defensive players are successfully kept away from the quarterback.

If all systems operated as expected then the play would have been completed with the desired outcome. Now what happens if the center does not successfully complete his task? The result would be what's known as a broken play and the device, or team, would break down, resulting in an unsuccessful outcome.

Let's go back and take a look at our car. It is much easier to understand how a car actually works once we understand that a car is comprised of numerous different systems. Each of these systems performs specific tasks, all the while working together, in unison, much like our football team. How many systems are there and what do they do? Here is a short list.

***Electrical:*** Provides electrical power to all other systems by means of a large battery. This system is key to operating the lights, starter, fuel system, and ignition system.

***Charging:*** Driven by the engine itself this system constantly charges the battery that powers the electrical system.

***Fuel:*** Uses power from the electrical system to deliver fuel from the fuel tank to the engine.

***Ignition:*** Uses power from the electrical system to ignite fuel delivered by the fuel system.

***Lubrication:*** Using a mechanical pump powered by the engine this system delivers oil to all moving parts inside the engine to reduce friction.

***Cooling:*** Using a mechanical pump driven by the engine this system pumps water or antifreeze through the engine to keep it running at an appropriate operating temperature. In some instances this system also uses power from the electrical system to operate cooling fans mounted to the radiator.

***Climate Control:*** Using power from the electrical system this system provides the passenger compartment with heated or cooled fresh air.

The list goes on and on but you get the idea. Notice how each individual system either provides resources to or requires them from another. No single system can perform all of the tasks required for the car to operate properly. In addition, if a single system malfunctions it will affect the overall operation of the car—sometimes with catastrophic results.

So why do you need to know all this? Suppose you are a mechanic and someone brings you a car that will not start. You get in and turn the key but you get no response. The starter doesn't engage and the engine doesn't turn over. What is the first thing that comes to mind that could be the cause of this problem? Since the car relies on the electrical system to operate the starter and the starter itself is not working when we turn the key, we need to begin our troubleshooting by checking to see if the electrical system is malfunctioning. How do we know what the electrical system is comprised of? By using the IPO method, we can determine the Input, Processing, and Output functions as well as the components that make up the electrical system.

Understanding that most devices are made up of numerous specialized systems performing specific tasks is key to really understanding how something operates. Without this knowledge, your ability to properly identify and resolve problems is severely hindered.

**Environmental Factors**

For the sake of this discussion let's define environmental factors as how the device affects or is affected by the immediate area to which it is confined.

So, what does that mean exactly? Let's say you are at home, it's a hot summer night, and you are running a

*Learn, Learn, Learn*

window air conditioner that is relatively old. All of a sudden the lights in the room flicker or dim momentarily. What do you think is happening? Is the power company experiencing a problem? Are you about to lose power all together? A short time goes by with no other issues, so you continue doing what you were doing. A little while later you notice the lights flicker again, only this time you hear the air conditioner momentarily make a loud humming noise at precisely the same time. What in the world is going on?

To explain this phenomenon let's take a quick look into the world of air conditioning. Air conditioners have a component in them called a compressor. This unit, driven by an electric motor, circulates the refrigerant necessary for heat exchange through the air conditioner's copper coils and also applies energy to the refrigerant. It runs in cycles and generally needs more electrical power upon initial startup than it does to continue to run through its cycle. When the air conditioner is new most of us would hardly hear when the compressor kicks in but as the unit gets older it operates less and less efficiently and therefore requires more power than normal to operate. This sudden need for more power, and the fact that window units tend to be connected to household circuits that are also in use by other devices, can put a strain on the circuit it is connected to. In more severe cases the entire electrical infrastructure of the house can be affected, causing lights throughout the house to momentarily dim.

Many other electrical devices can have the same effect on their environment. Appliances, power tools, you name it, can all have a negative effect on things like computers, TVs, and other sensitive electronic devices.

Once we understand how to apply the IPO approach to gain a solid working knowledge of how everything operates we will need to make sure to take any potential environmental factors into consideration. At that point, we can be confident that we can enter any situation and within a short period of time be able to assess it effectively. Then we can begin the process of isolating the root cause of the problem and conquering it.

## How is it built?

In a career as an IT consultant the greatest asset of all is a strong working knowledge of how different systems work and knowing their fundamental relationships to one another. I learned very early on that along with possessing a good working knowledge I also needed to have a good idea how things were actually manufactured. Understanding how a piece of equipment is put together can mean the difference between finding a solution quickly and forcing a customer to endure massive amounts of downtime. For example, it's not enough to know that all computers consist of similar components such as a motherboard, CPU, hard drive, RAM, power supply, and so on. You also must know whether or not those components can be interchanged with other computers or if they are specific to that particular model. All computers have a power supply but some computers use one that is specific to a particular model, thus making it more difficult to find a replacement should the need arise.

Another good piece of information I have learned over the years is how a network switch is constructed. For those readers who do not know what a network switch does, it is the device on a data network that connects computers, servers, and printers together. They are the backbone of the internet and they function in much the same way as an interstate highway system. A network switch allows data to flow from a source device to the destination device using a myriad of pathways. To effectively compare the two, imagine that the network cable (or your Wi-Fi connection) is the street you live on, with your computer acting as your home. This cable is plugged into a port on the switch, which could be considered the on/off-ramp to the highway. The backplane of the switch, which connects all the switch ports together, would be the highway itself.

Knowing how a switch is constructed became handy when the government agency I worked for experienced an outage that affected several mission critical machines (while I was on vacation, I might add). The switch appeared to be functioning just fine, as most devices connected to it were communicating on the network without issue. The machines in question were all connected to six adjacent ports and were the only ones not working. I moved one of the devices to an unused port elsewhere on the switch and it started working. I immediately knew at that point what I was dealing with. This particular switch was manufactured with a modular construction. The ports were installed in groups of six using modules that plug into the switch's backplane. As there were six devices not working and all plugged into adjacent ports on the switch, I was able to determine that they were all plugged into the same module and that therefore the module was probably defective.

Knowing that little tidbit of information was key to resolving the issue quickly, thus keeping downtime on mission critical systems to a minimum. To my coworkers that bit of information about how the switch was constructed was apparently inconsequential (and so was the fact that I was on vacation). In the end that one piece of information proved to be the missing link when trying to resolve this issue, thus proving once again that rule number two (you can never know too much about the products you are working with) remains true!

**Installation and Configuration**

The last bit of knowledge we need to add to our list is understanding how something is installed and/or configured. This can be for anything from a computer, with its software and peripherals, to household appliances and anything in between. The simple act of knowing how something was installed and how it is configured before an incident occurs can help us immensely when the time comes to resolve an issue. For example, knowing where the water line that connects to your dishwasher is and whether or not it has a shutoff valve installed where you can easily access it is crucial information to know. Especially if your dishwasher decides to shower you with more than a clean set of

*Learn, Learn, Learn*

dishes! If you have a new digital device of some sort, knowing how it is configured and, more importantly, knowing how to back up that configuration and the process to restore it are very important. We all know it doesn't take much for something bad to happen, resulting in the configuration being deleted or reset to factory defaults.

This may seem like a lot to learn but in actuality it is not. Most of this information can be obtained fairly quickly and easily. In doing so you will start to see patterns in how things operate, how they are made, and how they are installed. The basic principles—or once again, fundamental relationships—are all the same, but the application may be different from one thing to the next. The good news is the more you exercise these new skills the easier acquiring this information becomes.

# CHAPTER 4
# RESOLUTION SKILLS

NOW WOULD BE a good time to take a quick look at the overall process of resolving problems. In the previous chapters we learned that our most important weapon in the fight against downtime is to learn as much as we can about everything. So now that we have learned everything about everything our next objective is to learn how to apply this newfound knowledge should an actual incident arise. There are many schools of thought on the subject of problem resolution (or troubleshooting, as we call it in the service field), but the simplest and most effective method I have found is what I like to call "Learn, Identify, Isolate, and Conquer." Pretty self-explanatory, is it not? For most people who work in the services industry, the answer is yes, but you would be surprised at the number of people I meet who are unable to grasp this concept.

The most crucial step in resolving any issue is to properly identify it. This simply means that you need to understand *exactly* what is happening. Sounds easy, right? Sometimes it is. I mean, if you are a mechanic and someone brings you a car that is leaking fluids and

*Resolution Skills*

running extremely rough, and a quick visual inspection turns up a hole in the engine big enough to stick your hand inside, there is a pretty good chance you can rule out a broken belt as being the culprit. More often than not properly identifying the problem can be time consuming as well as confusing. Sure, anyone can see what is happening on the surface, but it takes the knowledge of what we learned in the previous chapter to really have an understanding of what is actually taking place. Knowing how something is *supposed* to work—and more importantly, how it operates—is the key to figuring out why it is no longer working. This is where our learning process pays off. Even if you are not the person who is eventually going to resolve the problem, doing your best to assist in identifying it now can lead to less downtime and a smaller repair bill later.

One thing you may notice throughout this book is that I try to avoid using the phrase, "What is wrong?" I will admit there will be times when I do for simplicity's sake but, in my opinion, this is a phrase that should be avoided as best you can. My reasoning is simple. When trying to identify a problem we are not looking for something that is wrong. We are trying to identify what is not working. I know, it sounds like the same thing, but it isn't. To a client, implying that something is currently wrong means there is a good chance that something was not done correctly from the start. This gives them a false impression that there was a problem with this device from the beginning and only now is this problem coming to light. In my experience, it is always better to ask, "What is it NOT doing?" instead. This implies that the device was just fine until something broke. Trust me.

*4 Steps to Solving Your Problem*

You will have far fewer issues with customers by avoiding the "W" word.

If you are a service provider you will also have to understand how to communicate with those whom you are helping. End users are not always the most informed on the operational characteristics of the things they use. They simply know how to do what they need to do and that is pretty much all they care to know. So, if the most important skill you need to develop is the ability to apply your working knowledge of the device and its environment, the second most important skill to develop is your ability to decipher what the end user is trying to tell you about the problem with said device. Trust me when I tell you this is not always very easy.

Every end user is different. They all do things their own way and they all have their own perspectives about how things operate and what is happening. Factor into that equation some people's inability to handle any unplanned event, such as an outage or failure of any sort, and the level of stress such an event can arouse, and you can immediately recognize the importance of understanding how to relate to your client in a calm, cool, and collected manner. Most of the time clients are capable of explaining the situation, but sometimes their descriptions are based more on their own theories than on actual reality. And let's face it, some are pretty creative in their ridiculousness. In fact, every example in this book, no matter how far-fetched it may sound, is based on something that actually happened to me or a close colleague.

*Resolution Skills*

What we all need to be mindful of is that each person you perform a service for is in possession of his or her own unique personality and associated quirks, just like you! Luckily for you these unique personalities exhibit a finite number of behavioral characteristics, which you can read about in the next chapter. Though not a definitive list of some of the more colorful citizens in our quaint little society, it does cover the personality traits you will experience the most as well as give you some insight and tips on how to effectively work with them.

The next step in this process is to isolate the problem. For the sake of this discussion we will define the word isolate as singling out a root cause. Again, this is where we can really use our newly obtained working knowledge. Because we know what something is supposed to do, how it does it, how it was put together, and how it was installed or configured, we can now use that information, in conjunction with what we have identified as the actual problem, to isolate the root cause. We will use the process of elimination to rule out any possible causes within the device itself, all the while paying strict attention to any environmental factors as well.

How can the local environment in which the device resides be a potential factor in its failure? Imagine for a moment you are operating a machine that uses an electric motor to perform its functions. Over the course of a few days you notice the machine is experiencing small problems throughout your shift that normally would not occur. With each passing day the problems

*4 Steps to Solving Your Problem*

seem to occur more and more frequently. Then, all of a sudden, the electric motor that drives the machine quits working. You had noticed a peculiar smell, and you saw a slight hint of smoke coming from the motor itself, but you never paid any attention to it. Thinking back, that was a sure sign that there was a problem. A service technician replaces the failed motor and gets you back up and running. But a few days later you notice that the machine is exhibiting the same behavior as it did previously, shortly before the last motor failed. And sure enough, a few days after the problems reoccurred, the new motor fails as well.

Now there could be any number of reasons why the replacement motor failed. Could the technician have replaced the failed motor with one that had a manufacturing defect? This would certainly explain the premature failure of the replacement motor. But what if there were an environmental factor to blame? Perhaps something outside of the motor itself was contributing to its failure. When the technician replaced the original motor did he test the electrical service that powered it to confirm it was within specifications? Did he do a visual inspection to determine if the motor was mounted correctly and aligned properly? Why would this be necessary, you ask? Because these could, in fact, be environmental factors, things outside the motor itself, which could have contributed to its premature demise. When trying to isolate the root cause of a device failure the key word to keep in mind is "root." Sometimes, what can seem like an obvious cause of your problem is actually just a symptom. Proper due diligence should be practiced at all times to insure you have eliminated any environmental factors and have, indeed, discovered the root cause of the failure.

Another thing we need to keep an eye out for is problems that are not really problems at all. In my experience, a great number of "problems" that are reported to me are not actual devices in need of repair but instead are issues related to training (or more to the point, lack of training). In this age of rapid change with an eye on increased productivity and a positive bottom line, it seems that this most important aspect of creating a solid and effective workforce is completely overlooked. It is frustrating to be called by a client to resolve a problem with something only to find that the real issue is that the user hasn't been properly trained in its use. Your first instinct may be to become angry or upset that you were called out to a false alarm, but I don't like to look at these situations in this manner. The best course of action is to work with the user and provide the necessary training, to the best of your ability, to ensure that the user can competently operate the device. I know it's not your job to train your client's employees, but a few minutes of your time now can save you many hours later. It's also not a bad idea to report this training issue to your primary contact for the client, as it may be something he or she is unaware of. Simply pointing it out to them can help reduce the number of training related calls as well as solidify a better relationship with the client.

Lastly, we need to conquer the problem. Since we know how it works and we have found out why it's not

*4 Steps to Solving Your Problem*

working the next step is obviously to find a solution. Again, sounds like common sense, right?

Over the course of my career I have been amazed, if not astonished, at some of the repairs I have seen people make. From printers being held together with surgical tape (yes, you read that correctly) to broken serpentine belts on a car engine being spliced back together with heavy duty staples (you should've seen the damage that little "repair" caused), the creativity employed by some in the repair industry, however misguided, is second to none.

It is extremely important to have a strong working knowledge of what you are working with in order to employ the appropriate repair. The solution for your problem could be as easy as making a quick configuration change or replacing a part. Depending on the nature of the failure, it could entail a complete overhaul of how it was installed or configured or modifications to its environment or support structure (or, even worse, how it was manufactured).

A very important aspect of determining the proper course of repair is to take into consideration the environment in which the device resides as well as how it is used in that environment. The worst thing we can do is to provide a solution for a light duty environment when an industrial solution is required. Conversely, you do not need to over-engineer a solution for something very minor in scope. If a 20-year-old plastic part has finally worn out chances are good that a case-hardened steel replacement part would be unnecessary. Also,

*Resolution Skills*

always remember that duct tape is a wonderful invention but just because it *can* hold a machine together doesn't mean it *should* be used to hold a machine together! Ingenuity is great, but common sense works best.

This may sound like a difficult process. And while I agree that it can seem somewhat overwhelming, I can assure you that once you start to use this method it will change the way you troubleshoot and resolve problems. In addition, your ability to manage stressful situations with clarity and focus, whether in your personal or professional life, will greatly improve.

# CHAPTER 5
# PEOPLE

PEOPLE. THEY CAN be your best tool in solving a problem or they can be the biggest barrier. To a service provider the mere mention of the word "people" can cause headaches and undue stress, and the word itself is sometimes used in a rather derogatory fashion. If I were to poll 100 service techs and ask them to define the majority of the people they come into contact with during a normal day that definition would surely include the words panicky, confused, and, the one that makes me laugh the most, bastions of misinformation.

Numerous acronyms and phrases have been used by service personnel over the years to describe their various clientele and the mischief they have been known to cause. Acronyms like the good old PEBKAC immediately come to mind. PEBKAC stands for Problem Exists Between Keyboard And Chair. Then there is this oldie but goodie, the I-D-10-T error. Look closely and you will see that it actually spells IDIOT. A good friend of mine has been known to inform certain clueless clients that the problem was simply a loose nut behind the keyboard. Think about that one for a minute.

*People*

How many times have you been in a situation where something is wrong and no one knows what is happening? No one can answer a simple question as to what is happening and meanwhile, everything that should be getting done is at a complete standstill. These are the types of situations I walk into every day. What makes this worse is that the very people you are there to help can actually hinder your ability to help them. They accomplish this by not giving you the information you so desperately need. Would you like an example? How about that lovely little customer who complains his new computer will not turn on and after 15 minutes of troubleshooting over the phone, you hear him ask a colleague, "Have they turned the power back on yet?" Hmm… I think I may have found your problem!

Here is my personal favorite. You ask someone what is wrong and the answer is "nothing is working." Well, define "nothing" for me, please. The office lights are on so I know the building has power. I see people using the copier so I'm pretty sure that is working also. I see lots of things that are functioning normally. So please tell me, what exactly isn't working? Instead of being told in a coherent manner the depth of the problem at hand, I am on the receiving end of a barrage of meaningless and disconnected details. Before you know it I feel like I know less now than when I walked in and that I am in possession of fewer IQ points than what I arrived with.

Some people just do not possess the ability to examine a situation in a logical and coherent manner and then describe what they are experiencing. And although this can be very frustrating for you to deal with it is necessary that you understand it. End users are normally trained in how to use something and not how it operates. Oftentimes they are also not trained in how to fix it once it stops working. Most end users do not take the time not only to learn how to operate the device but also to learn how the device operates. How you operate a device and how a device operates are not synonymous with one another. Knowing how the device operates can be useful information that can be used to perform the job more efficiently.

Want proof? I have worked for an individual who, on occasion, would need to forward an attachment they received in their email to everyone at their office. Instead of saving the attachment on their computer and attaching it to a new email they would actually print the attachment, go down the hall to the copier, scan the document back into the computer, and then attach it to a new email. The idea of just saving the file and then attaching it instead of adding the steps of printing and scanning it back in seemed foreign to them. To this day I cannot get through to them that there is a better and easier way.

What you as a service provider need to accomplish, first and foremost, is to learn how to deal with the various people and their associated personalities (or lack thereof) who will be relying on you to resolve their problems. Identifying and understanding the personality of the person you are trying to help is almost as important as identifying the actual problem you are there to resolve. Perhaps the most important skill you must develop is the ability to use the subtle tactics needed to

*People*

extract the information you desire from those who wish to help (and sometimes from those who wish to hinder)... and you need to do this in a manner that elicits their confidence in your ability to fix their problem. In some cases, you may even need to instill in them the confidence they need to be able to help you.

And you thought you were just going to waltz in there and fix a problem? So, without any further ado, let's take a look at some of our clientele.

**The Genuine Helper**

This person is somewhat of a rarity. They take their job seriously and learn, to the best of their ability, as much as they can about the devices they use. They take great pride in being able to solve simple problems without the need to call for help. This person will go above and beyond to help you and will be able to provide you with not only the information you need about the problem at hand but also with an accurate history of issues including how they resolved them. This can be very helpful when looking for problems such as installation or configuration issues, failure patterns with the device, and any possible trends with other similar devices within the environment (more on those later). This person will also be the first one to speak up if they are the one who caused the problem, and they are usually very anxious to learn from their mistakes. This is the one person you can speak to on a somewhat technical level as long as you do not confuse them with a lot of industry jargon. Simply explaining to them what you need, what you found, and how you resolved it goes a long way with this person should you ever be called back again.

## The Honestly Ignorant

"I know just enough to be dangerous" is perhaps the most used sentence by this person! The Honestly Ignorant person is somewhat of a rarity, much like the Genuine Helper, but the moniker is a bit of a misnomer. The word ignorant doesn't truly characterize them, because they will be the first to tell you that they have no idea what is wrong and that they know just enough to do their job and that's it. Anyone who can freely admit that they have just enough knowledge to be dangerous is a lot smarter than they give themselves credit for. They have the ability to let you know exactly what part of their process is not working and with more accuracy than you might expect. Sometimes that's all the information you need to be successful!

A more fitting description of this person would be "ignorant by choice." It's not that they lack intelligence or are unable to comprehend what is going on, but instead they lack the confidence to delve deeper into the process. They are afraid they will not be able to understand something or, even worse, that they will break something while doing so. You can also guarantee that if this person calls you with a problem more often than not it's a real problem. Chances are good they are not going to do something they are not supposed to do (again, for fear of breaking something). When talking to this person you MUST speak clearly and deliberately, using plain and simple language. Walk through their issue step by step, letting them lead you to the information you need instead of pushing them to it. If you push too hard they may become frustrated, confused, or even worse, their fears of being labeled as incompetent will take over. If that happens, rest assured, you will be on your own. If you need clarification on something they are trying to explain to you be sure to ask as soon as the question arises and not after they have finished. That's when confusion can set in for both of you and, again, you will lose them. They can, without realizing it, provide you with more than enough information to assist you as long as you are calm and direct and instill confidence in them.

**The Helpee Helper**

The Helpee Helper is anything but helpful. They mean well but that doesn't translate into meaningful assistance during a time of crisis. They usually have no idea what's going on, how anything works, or how to obtain the information you need. What they do know is that something has gone horribly wrong and they want to be in the middle of it. They will offer to help but in all honesty they are more of a hindrance than an asset. Politely decline their assistance and move on. A simple comment such as "I think I have everything I need" or "Who can I talk to that is actually experiencing the problem?" will usually send them on their merry way. You will be better off finding the problem on your own and will probably resolve it much quicker.

## The Know-It-All

How many of us know this person? This is the person who believes that just because he knows how to do one thing he can do everything else as well. There is nothing he can't do, nothing he has never done, and, more importantly, he has done it better than most people, including you. In reality, this person is most likely intimidated by your presence because chances are quite good that he caused the problem you are there to fix. His ego, however, will not let him admit it. Because of this he can sometimes be difficult to get on task. He will tend to spend time trying to dazzle you with his brilliance, all the while minimizing his involvement, instead of giving you the information needed to expedite a solution.

When this occurs, I've found that politely and firmly explaining exactly what I need and nothing more is the best way to help this person gain focus. Make it clear that you are in charge and that you need detailed information from him in order to resolve the problem. Make your queries very specific, such as "Please explain to me exactly what happened immediately before the failure." Let him know that you don't care who created the problem but you need to know what was done. If he veers off subject make it a point to lead him back in the right direction. Maybe ask if he was the one using the device when the failure occurred. That usually puts the ego in check and at the same time puts him in a position to help. His fear of possible embarrassment if he is, indeed, responsible for the failure is understandable. Affirm to him that there is no need to be embarrassed if he did something he shouldn't have. After all, you've seen it all before. There have been quite a few instances where I have used this approach, and whenever I have been called back to that particular client for other issues this person has been the most eager to assist and learn.

## The Expert

While the Know-It-All claims to be all-knowing as well as being better at anything and everything than anyone else, the Expert claims to know absolutely everything about his job and yours. Whereas the Know-It-All likes to brag, the Expert likes to belittle and will go out of his way to goad you into a conversation where he can attempt to dazzle you and everyone around you with his cynical wit and intelligence while attempting to make you feel inferior. Most people are annoyed by the Expert but I am actually quite entertained by them and do so enjoy having fun at their expense. (Have I mentioned that I can be a bit cantankerous at times?) It's one thing if they really do know what's going on and are willing to assist you in any way possible, but usually they seem to get their jollies by trying to make you feel as though you are somehow beneath their contempt.

I find it best to distance myself from these folks as soon as they are no longer entertaining to me, but as they tend to be clingy and have an insatiable craving for attention, sometimes you just have to give them that which they desire. Simply asking them what they think is the problem is usually all it takes to divert their attention elsewhere. Chances are they have no idea what the actual issue is, and politely asking for their help puts them in a precarious situation. They either answer incorrectly, blowing their cover as the "go-to" person for all things job related, or they simply respond that they don't know. Either way it will not be long before they find something else to do. If they do appear to be eager to help, always be polite and to the point, and let them know ahead of time that they need to do exactly as you say, when you say it. This person may have the tendency to try and exhibit their intelligence by anticipating your next move or request, usually at the most inopportune time, potentially creating a much bigger problem.

**The Blame Seeker**

This person's sole purpose in life is to find something wrong and then search out some poor hapless soul to pin it on. They are more worried about who created the problem than they are about trying to resolve it. If you happen to run across a Blame Seeker you need to do your best to divert their focus to the problem at hand. The best way I have found to accomplish this is to remind the Blame Seeker that it is not my concern who did it and the more time we spend trying to find out who created the problem the longer it will take to fix it—and, more importantly, the more it will cost. They must realize that the main objective here is to fix what's broken.

*People*

## The Dribbler

As loveable, kind, and gentle as this person appears to be, they can be the most frustrating to deal with. That is because they know exactly what happened to cause the problem you are working to solve but they apparently have trouble giving you all of that information up front. Instead you get bits and pieces of information dribbled to you a little at a time and in no particular order. Usually the most crucial piece of information, the piece that ties everything together, is the last bit of info that slips out, usually prefaced by the phrase "oh, did I mention that…". This usually happens because the Dribbler is actually responsible for creating the problem but is simply too embarrassed to admit it. They feel that we will think they are stupid if they admit what they did. You can always spot them, too. When they are describing the problem, they tend to be very quiet and timid. They appear almost childlike in demeanor and refuse to look at you eye to eye.

When you see this behavior remember the last thing you want to do is further their embarrassment. You need to gain their trust by assuring them that whatever they did is no big deal. You've seen it all and probably have done worse yourself. We all have! Heck, I've unknowingly tied a network cable into my shoelaces and then walked away. In the process, I pulled the cable loose and brought down a substantial piece of a mission critical computer network. It happens to us all and they need to know that.

Most of the end users you will encounter will be normal everyday folks. When they run into problems they just want someone to be able to fix it as soon as possible and with little to no stress. And while the examples and descriptions I have provided for you may seem a bit extreme I can assure you that these people do exist. I've met them and sometimes they make you want to go into a room and scream. But the one thing you must always keep in mind is that you are there to help, and you convey that message by the way you relate to your clients. In some ways you are more than just a service provider. Sometimes you are also a psychologist, confidante, and friend.

# CHAPTER 6
# IDENTIFY THE SITUATION

YOU HAVE LEARNED that the best way to be prepared for any crisis you may face at your workplace or at home is to have a good working knowledge of the things around you. The next thing you need to be able to do is to use that knowledge to identify exactly what is going on should a problem arise. In this chapter, we will take a look at a number of examples of how to properly identify what is actually happening in a given situation as well as some ways to identify a potential problem before it becomes a real one.

You will notice throughout the next few chapters that I place a great deal of emphasis on asking and answering questions. This isn't just something I am doing to help lead you through the scenarios we are studying. Instead, this is a habit I have formed over the years that helps me immensely. Whenever I am presented with a problem the first thing I do is to observe what is happening and then start asking myself questions about it. In my mind, the goal is to find the answer to those questions, which will, in turn, allow me to properly identify what is taking place. Doing this not only helps me to navigate my way

*Identify The Situation*

from point A to point B but it also allows me to further bolster my working knowledge of the device. In a lot of instances, I have learned more about things by trying to fix them than I did while initially learning about them.

## Four Out of Five Senses Agree

First, let's answer this question. What is the difference between being a good service technician and a great one? Answer: A good technician can fix whatever is broken in a timely manner, keeping downtime and disruption to the client at a minimum. A great technician can quickly identify an existing problem (and sometimes even a potential problem, before it occurs) and create a plan of action to resolve it with little or no downtime or disruption to the client. Remember reading in Chapter 3 about the importance of learning about the things around you? You also discovered that it is important to watch and listen intently to what a device is doing and the noises it generates under normal operation. Possessing this knowledge of what a device does during normal operation gives us an immediate advantage when trying to identify the current problem. Coupling this knowledge with a keen sense of observation can also help you to identify potential problems before they occur. This allows you to alert your client to the existence of a potential problem and gives you the opportunity to present resolution options in advance, thus minimizing or eliminating costly downtime.

Properly developing your observational skills is actually easier than it sounds. Even someone who is known for not noticing the blatantly obvious can develop effective observational skills with a little practice and persistence. It just takes time.

So how do we develop these skills? First, we need to identify the kinds of things to look out for. Problems—or potential problems—can identify themselves in any number of ways but they can be grouped into four basic categories. There are visible problems, sounds, smells, and vibrations. By employing our senses of vision, hearing, smell, and touch, and combining them with our working knowledge of our surroundings, we can greatly increase the speed at which we properly identify an existing or potential problem. Notice, if you will, I only mentioned the use of four of our five senses. Unless you are troubleshooting a problem with a chef I don't really think tasting things will help you much in your career as a service technician. But, then again, I have been wrong in the past.

What are we looking for, you ask? Everything! Thinking back to our earlier example involving the failure of an electric motor, you will remember that I mentioned that the user noticed the presence of a peculiar odor and the slight hint of smoke before the motor failed. I also wondered, while performing the repair, if the technician noticed anything out of the ordinary, like a possible alignment problem with the machine or if he had tested the electrical service to the motor. Had the user, upon recognizing the odor emanating from the electric motor, alerted someone, they could have investigated the matter, identified the problem or potential problem, and

*Identify The Situation*

then taken the necessary steps to resolve any issues before a failure occurred.

One habit I have created for myself throughout the course of my career, especially when visiting a new client, is simply to observe everything. I want to know where things are, how things are connected, and details of the environment in which they reside. I try to listen to each piece of equipment during normal operation and compare how it sounds with my knowledge of how similar devices sound. I also try to get a feel for the overall environment, its noises, smells, and any other traits it possesses. I truly believe that having this information gives me numerous advantages when the time comes to identify a problem.

Let's take a look at some examples of how simply being observant and aware of your surroundings can assist you in identifying a problem. If you are like me, whenever you go out to your car the first thing you do is give it a quick look to see if anything has happened to it. You look at the windows to see if they are dirty or damaged. You glance at all of the body panels and paint to see if there are any new dents or scratches. You even take a look at your lights to make sure they aren't broken.

Let's say you look at your tires and immediately notice you have one that appears to be going flat. A quick visual examination, as well as the application of the old push on the sidewall, indicates to you that the tire is low on air pressure. You look in your toolbox or dig through your glove compartment and pull out your handy-dandy tire pressure gauge to see if your suspicions are correct.

*4 Steps to Solving Your Problem*

You unscrew the cap from the valve stem, apply the tire pressure gauge, and sure enough that tire is about 10 pounds low. A simple glance at your tire quickly informed you of a potential problem! Now what do you do? If you are lucky enough to have an air compressor at home you could pump it up right there. If not, you could drive up to the nearest gas station and use their air pump. That would fix it.

But the question that should now be running through your head is WHY the tire was low in the first place. So, what's your next step? Well, if you took the time to learn a little about how to maintain your car you know that tires periodically need air. If it has been warm outside and then the season changes from summer to fall and temperatures drop, then the air pressure in all of your tires will drop also. In this scenario, changes in the environment caused the air pressure to drop, and to resolve the problem you simply need to add more air. So, your next step is to check the air pressure in your other three tires. If they are all down about the same amount then you have identified the problem as environmental. You can now take the necessary steps to resolve the issue by adding the appropriate amount of air to all of your tires.

But a quick look at the remaining tires shows that they seem to be properly inflated. The tire pressure gauge confirms your observation. Now what do you do? Well, ask yourself what could cause this tire to be low but not the others? Is there something stuck in the tire, causing a leak? Could it be a nail, perhaps? How about a bent rim? Maybe the valve stem where you put the air in the tire is

*Identify The Situation*

leaking. So now we have to check the tire closely for any signs of damage that could be letting air escape. Do you see any damage to the rim? How about the valve stem? Do you see a nail or screw stuck in the tread or sidewall of the tire? If you find any kind of damage, then congratulations! You have now potentially identified the problem.

If you don't find anything, no worries. All is not lost. Depending on your knowledge of cars and your comfort level for moving forward you can now decide if you want to further troubleshoot and isolate the problem or take it to someone who may be better equipped for the job. But what you HAVE done is narrow down the problem, thereby giving the mechanic much-needed information to help find the exact cause of the leak. And it all started by knowing what the tires are supposed to look like under normal operating conditions and by observing a noticeable difference.

Now was that difficult? No, it wasn't! Here is another example. A customer calls me and says he cannot print to a printer in his office. Having made it a point to take mental notes of each of my client's environments, I know that the printer he is referring to is a shared printer connected to the office network and that everyone in the office has permission to print to it. I begin to ask him a series of questions related to the problem, starting with very general inquiries regarding the environment. I ask if anything has changed, whether others can print to the printer, if the printer is displaying any error messages, etc. During our conversation, I quickly determine that he is the only one who cannot print to this particular

*4 Steps to Solving Your Problem*

printer, so my focus now turns to his computer. By knowing the client's environment and how the printer was installed and configured I was able to quickly determine where the problem was centered, saving me—and the client—a great amount of time.

There is an art to properly and easily identifying a problem. Fortunately for all of us, it is an art we all have the ability to learn. Knowing how to analyze a situation by using your observations and your working knowledge of the device (as well as the environment in which the device resides) helps you narrow your search for a root cause. It is at this point that you can begin the vital task of isolating the root cause and then creating a proper solution.

What if you don't have any knowledge of the environment? What do you do if you are thrust into a situation where no one knows what is going on and everyone is looking to you for answers? Don't worry, because I've been there, too! In this case, the first thing you need to do is think back to our lesson in Chapter 3 on the importance of learning everything about everything. You would use the IPO approach to identify how the device in question operates and then try to find similarities with things you already have knowledge of.

Let's say you are faced with trying to identify a problem with a clothes washing machine. You know nothing about them but you have some experience with a dishwasher. Well, what does a dishwasher do? Isn't that similar to what a clothes washer does?

*Identify The Situation*

Water is introduced into the dishwasher by an electric pump, an electric motor is used to rotate the spray arms that spray water onto the dishes, detergent is dispensed, and then, when the dishes are clean, the water is pumped out. Similar to the dishwasher, the clothes washing machine uses an electric pump to introduce water into the wash basin, and an electric motor spins the basin to agitate the clothes in the soapy water and also to spin dry the clothes between wash and rinse cycles. Then another pump removes the dirty water from the machine. Now that we have identified the similarities we have to find out how the clothes washer performs those similar tasks differently from the dishwasher, and along the way try to identify exactly what is not working.

Sounds like common sense, doesn't it? Well it is, actually. The problem is a lot of people do not understand that these similarities exist. They see a dishwasher as a dishwasher and fail to understand that it not only performs a similar function but also operates in a similar fashion as a clothes washer.

If you go back to Chapter 3 and reread the segment on Input, Processing, and Output and apply it to our dishwasher-clothes washer comparison you will see each step in the IPO operation.

- Input - Water is introduced into the unit with the dirty dishes or clothes

- Processing - Motors drive the process of cleaning the dishes or clothes by rotating spray arms or a wash basin

*4 Steps to Solving Your Problem*

- Output - Waste water is pumped out of the unit at the end of the cycle, resulting in clean dishes or clothes

Now I know this seems like an overly simplified way of describing the operation of two somewhat different and complex machines. We know there is a lot going on inside that metal box than what I've listed here. But before we can begin to successfully identify a problem with a device we must first understand what it does and the process it uses to accomplish its task. Knowing this basic information allows us to look deeper into the device, identify the systems in use for each process, determine the components needed by each system to perform the assigned task, and, finally, pinpoint where the operating process is failing.

By always applying the Input, Processing, and Output methodology of defining the operating process of a device we are able to accomplish many things. First, we can take our working knowledge of the products we use and directly apply it to devices of a similar nature. Next, should problems arise, we can use that same information to identify where in the operating process the failure is occurring. And finally, we can be confident that if we enter a situation where we are being asked to identify or resolve an issue with a product that is completely foreign to us, we can employ this process to quickly identify similarities with devices we are already familiar with.

# CHAPTER 7
# ISOLATE THE PROBLEM

ONCE WE HAVE identified what the problem is our next step is to isolate the root cause. Sometimes this can be easy and other times it can be quite the daunting task. It is at this point where everything we have learned thus far can actually be employed. All of our studying and learning about how things work, how they appear when operational, and the affect the immediate environment has on them will need to be incorporated into a cohesive mesh of information in our minds that we can access on a moment's notice.

Okay, I'll admit that was a little overly dramatic but, for the most part, entirely true. There are many schools of thought regarding the "proper" way to isolate a problem and, in my opinion, they all work. The real question is which method is most efficient and, more to the point, most effective? Each person has a different way to broach this subject, but since I'm writing this book, we will focus on the method I use.

Over the years I have found that the easiest way for me to isolate the root cause of a problem is by employing what I like to call the Outside - In approach. Very

*Isolate The Problem*

simply stated it is a means of eliminating possible causes by looking at the most general and obvious possibilities (Outside), ruling them out based on the information I have obtained about the problem, and then working my way down to the more specific (In) possibilities. It's a method similar to what police detectives use in that you narrow your search by ruling out possibilities based on the immediate evidence you have available. Along the way you will be presented with more information, or clues, eventually leading you directly to the source.

What's that? You would like an example? Okay. Let's go back to something we touched on earlier. Let's say you are a mechanic and someone brings you an older model car that has fluids leaking out of it, smoke billowing everywhere, and it's running extremely rough. What is your first step? First, identify exactly what is happening. We know, based on our automotive knowledge, that a car is not supposed to be blowing smoke or leaking fluids and should be running much smoother than this one is. So, we need to observe exactly what this car is doing and answer a few questions that will help us identify what is going on. With that in mind, what color is the smoke and where is it coming from? Is it coming from under the hood or from the exhaust pipe? If you know anything about cars then you know that bluish-white smoke is usually produced from burning engine oil either from a leak in the engine compartment or from the exhaust pipe. Black smoke is usually found exiting the exhaust pipe and is usually caused by the engine's being fed an improper fuel mixture or more fuel than it can burn. This is also

*4 Steps to Solving Your Problem*

known in the automotive world as "running rich." The smoke we are experiencing is of the bluish-white variety and it is coming from under the hood. So now we know that we have smoke from burning oil coming from under the hood and we also know that there is some sort of fluid (we can assume it's oil) leaking from under the hood. If we take a quick glance under the car we can see that there is, indeed, oil dripping as well as some sort of green fluid. Upon further inspection, we determine the green fluid to be antifreeze.

Now that we know exactly what is happening we will use that information to search for and isolate a root cause. We open the hood and look around the engine compartment for any broken or loose hoses but find none. Next, we look at the top of the engine and start working our way down to see if there are any leaks emanating from any gaskets. For those of you who are unfamiliar with how an engine is assembled, gaskets are used to seal any opening where two separate engine components meet to insure that any fluids that travel between the two do not leak out. For example, the cylinder head and the main engine block are separated by what's known as the head gasket. Upon close inspection, we do not find any evidence of a gasket leak. As we look deeper, though, we can finally see where the fluid is leaking from, and thus, the root cause of our problem. There is a hole in the side of the engine block!

I used this example to serve two purposes. First, it is somewhat of a metaphoric example, as we started our search for a cause of this problem outside the car itself and worked our way into the engine compartment to

*Isolate The Problem*

narrow down the specific problem. More importantly, we demonstrated how having a thorough working knowledge of things can be a valuable tool to effectively get to the root of the issue. If you had not acquired the basic knowledge needed to answer those initial questions it would have been much more difficult to find the source of the problem.

As I mentioned in the previous chapter, I place a great deal of emphasis on asking and answering questions, both of myself and especially of others. For me it's the best way to keep myself on task and it makes it easier to work my way through a situation. In fact, whenever I seem to be having difficulty in determining the cause of a problem I take a step back, reevaluate the situation, and then attempt to answer my initial questions again. When I do that I seem to be better equipped to find the correct answers and, in turn, find the cause of the problem.

In the previous example I made sure to answer the questions about the origin and color of the smoke coming from the car as well as the types of fluids leaking from it. At first, those questions may have seemed somewhat insignificant, but as we progressed we learned that the answers to those questions played a key role in identifying the problem. In the remaining scenarios presented to you in this book pay strict attention to the initial description of the problem and the questions that are asked about the observations being made. Then see where the answers to those questions lead you.

## Questioning Change

Perhaps the most important question you will need to answer when trying to isolate a cause is this. What has changed? Change is inevitable but sometimes it's accompanied by unintended consequences! A good rule of thumb is to always remember that if something was working yesterday but it's not working today there is a good chance that something changed. The real question is whether or not the change took place in the environment or within the device itself. If you ask someone, "What has changed?" invariably you will receive the same answer I receive on almost a daily basis. That answer is usually, "Well, nothing has changed." To quote my good friend Rebel Phillips, "Sometimes I feel like saying, well, you and I both know that fundamentally something HAS changed because something that WAS working ISN'T working now! I'd say that's a rather substantial change. Wouldn't you?" (Ah, the sarcasm runs deep with this one… I trained him well!)

In the absence of any evidence showing us what, if anything, has changed, we need to do two things. First, we need to try to identify any differences in the environment since our last visit. (Obviously, if this is your first visit to this particular client, you can skip this step.) Let's say the complaint is about a copier that is all of a sudden not working correctly. You might want to examine the environment in which it resides. If you notice a 30-year-old refrigerator and an electric space heater have been moved into the room with the copier—all plugged in to the same electrical circuit, I might

*Isolate The Problem*

add—then chances are pretty good that you have found not only what has changed, but also the source of their problem.

The second thing you need to confirm is that no changes have been made to the device itself. Usually a quick visual inspection can determine if any modifications have been made. If it is an electronic device that uses software, or firmware, to configure how it functions, then make sure to look at its configuration to see if anything obvious catches your eye. If you have a saved or printed copy of the configuration then this would be a good time to do some comparisons. I guess what I am saying here is never trust anyone who says nothing has changed. Chances are good they are wrong.

Here is another wonderful example of the incorrect use of the phrase "nothing has changed." I was visiting a new client for the first time. The complaint was that their server would stop responding numerous times throughout the day. I asked if there had been any changes made to the server or the building recently and, of course, the answer was no. I was led to a closet where the server resided. Notice I said "closet." It was a broom closet, to be exact. I opened the door to the closet and was attacked by what could only be described as the purest definition of heat itself. Here was a server with four processors, six hard drives, and two power supplies, all very proficient in the art of generating heat, all working together to not only serve data to all the users in the office, but also to generate enough heat to set the very room in which it resided on fire.

*4 Steps to Solving Your Problem*

My next question to the client was, obviously, "Has the server always been in this room?"

Her reply was astounding. She actually said, "No, it hasn't."

"So, when did you move it to this closet?" I asked.

"Oh, about a week ago," was her reply.

Being the inquisitive type I pressed on. "Isn't that when you told me all the problems started?"

"Come to think of it, you're right," she said. Her face lit up as if she were experiencing some sort of epiphany.

"So when I asked if anything related to this server had changed, the answer should've been 'yes,' right? I mean, moving it to a new location can be considered a change, can't it? Even if it's just a change of scenery?"

The brightness in her eyes immediately turned dull. "You would be correct," she said.

In this scenario, the cause of the server problems had nothing to do with the server at all. It had simply been moved to a new location that had practically zero ventilation, causing it to overheat. Needless to say, we soon found a new location that was better suited for housing sensitive electronic equipment.

*Isolate The Problem*

## Answer Me This

Let's get back to the task of asking ourselves questions by taking another look at our leaky tire problem. We determined that the tire had a leak, but when we examined it for any signs of puncture we didn't find anything obvious. Remember our little lesson on taking a step back and reevaluating our situation? It is at this point that we must devise a new plan to allow us to search for the exact source of the leak. We can do this by answering a few common-sense questions. So where do we start? Think about this for a moment. You know that there is a leak in the tire, so now you should ask yourself this question: What was used to inflate the tire? The answer is, of course, air. So obviously it is air that is leaking out of it. We can't actually see air but we know that air is, indeed, escaping, so we ask ourselves this: What can we do to help us "see" the air? If only there were a way to capture the air as it escapes from the tire. What is one thing you can think of that would help us to see the air escaping from this tire? Have you ever seen air escaping from anything before? I see it all the time in the form of bubbles!

Remember blowing bubbles when you were a kid? You dipped the little plastic wand into the bubble container and then gently blew into it to form cute little bubbles that would float effortlessly all over the yard. Or maybe you are old like me and remember the "Lawrence Welk Show" on TV. That man really enjoyed his bubbles!

So how do we turn this tire into a bubble machine? Well, first, we grab a bowl or a bucket and mix up a nice

*4 Steps to Solving Your Problem*

mixture of warm soapy water. Next, we slowly pour the soapy water onto the tire, making sure to cover the tread and sidewalls as well as the valve stem (the part where you put the air in). Then we just wait and watch. Before long any air escaping from that tire will be fairly evident, as you will start to see bubbles forming at the site of the leak. If you don't see anything after a minute or so simply add some air to the tire. The excess air pressure will push its way out of the troublesome hole. Once we find the leak we mark it and get it to a tire mechanic or, if we are so mechanically inclined, break out the tire patch kit and fix it ourselves.

Wasn't that sensible? By simply taking a step back, reevaluating the situation, and asking ourselves a series of questions related to the issue at hand we were able to devise a simple plan of action. The plan involved using soapy water to create bubbles at the precise location where air was escaping from the tire.

Let's revisit our printer problem from the previous chapter. By using our knowledge of the environment, knowing how the printer was installed and configured, and asking the correct questions of the person who was using it, we were able to determine that the printer was actually working just fine. It was our user who was apparently unable to print to it. Now we start eliminating possible causes. We start with the most general and simplest of possibilities and work our way down to more specific items until we find the actual cause. First, we ask ourselves if his computer is set to use this printer and if it is the default printer. Once we determine that it is, we check the program he is printing from. It is

*Isolate The Problem*

entirely possible that he was unknowingly printing to another printer that was assigned within the program itself rather than to the computer's assigned default printer. In this case he wasn't. Now we dig a little deeper. We look at how the printer is installed to his computer. We notice that it is using the correct software but that it is an outdated version. Chances are the printing software is either no longer functioning properly or is unable to support the printer itself, so we download the latest version and install it. Voila! The printer is now working!

**But what if...?**

Let's shift gears for a moment. Let's say you have followed all of the steps so far and have examined the issue in great detail but you still have not successfully isolated the problem. As is often the case, you simply cannot find anything that blatantly says, "I'm broken! Please fix me!" Remember our little discussion on the importance of examining the environment?

Not long ago I was called to a customer's office to check out a computer that was having problems starting up. Over the course of the last several months the computer's performance had dropped significantly. Each day it was generating more and more random errors until finally it just would not boot. I then remembered that this was the third time in just over a year that this computer had experienced the exact same problem (The previous times this had occurred the user had brought the computer to me but this time I went to her office to

*4 Steps to Solving Your Problem*

check it out). I ran diagnostics on the machine and determined that all of the hardware components inside were just fine, yet it still would not boot correctly. I reinstalled the operating system and all software on the computer once again—but then I noticed something in the corner. In the office with this computer was a small, rather old, dorm-style refrigerator. I reached into my toolkit and grabbed my multimeter. I stuck the probes into an electrical outlet and noted both the voltage and current levels. I then unplugged the refrigerator for a few minutes and plugged it back in. As soon as the refrigerator came to life and its compressor kicked on, I noticed the voltage level drop significantly on my meter. I did this a few more times to verify that this occurred each time the refrigerator's compressor came on.

So, what did that tell me? A lot, actually! Much like the engine in a car, a computer's hard drive (the component where your operating system, software, and data reside) relies heavily on timing to function correctly. A car engine uses a system of timing that allows the crankshaft, camshaft, and ignition system to work in unison. This is to ensure that the fuel that enters the cylinders is ignited at precisely the correct time. The ability to set the correct timing allows the engine to perform the task of igniting fuel over and over again with extreme precision, thus allowing the engine to operate at its maximum capability as well as to maximize its fuel efficiency. A computer hard drive uses the same theory to function. Whereas the car engine is designed to be able to maintain the proper timing at any speed, a computer's hard drive is not. The platter in which the data is stored spins at a predetermined and

*Isolate The Problem*

precise speed while the armature (the component that reads and writes the data) moves back and forth across the platter, also at a predetermined and precise speed. The platters and the armature are powered by separate electric motors, each of which reacts differently to identical drops in voltage. If there is a small drop in the voltage available to the computer, even for a split second, such as when the refrigerator compressor activates, it can greatly affect the functions of the drive. If one component of that drive slows down while writing data, even by just the tiniest fraction, the data can be written to one place on the drive but be reported to be in another. Over time, if this is occurring on a regular basis, the computer will experience a decrease in performance because the hard drive is unable to find the data it is looking for. Instead the drive spends its time retrying the read operations, eventually leading to errors and then, subsequently, failure.

However unrelated to the actual problem it may seem, my basic understanding of how a refrigerator works—knowing that when the compressor comes on it can draw more current than normal, and also knowing that the older the compressor the more current it can draw—was instrumental in helping me find the true cause of the problem at hand. You might think that this was one of those once in a lifetime events, but in all honesty this wasn't the first time I ran into such a problem.

Isolating a problem can be challenging, but armed with our knowledge of the environment we can use our outside-in approach and the process of elimination to isolate the root cause. By observing the situation and

*4 Steps to Solving Your Problem*

asking and answering key questions related to what we are seeing, we can efficiently isolate the cause and then begin to build a plan to resolve the problem. When an obvious cause cannot be found we must remember to take a step back and reexamine what is happening. If we take the information we now have and revisit our earlier questions we may find that our answers will lead us in a new direction. Lastly, we must remember that the environment in which the problem occurred can be a huge contributing factor to the current situation. Understanding and taking into account all of the factors related to the issue at hand is the key to successfully isolating its root cause.

# CHAPTER 8
# CONQUER

SO NOW WE come to the final step in our process. This is the part where we can actually fix that which ails us. We have learned everything about everything, we have identified the source of our grief, we have isolated the actual cause of said grief, and now it's time to fix it. So what do we do? We fix it, of course. But wait! Before we do that, there are some things we need to keep in mind. Most people would assume that since they have finally nailed down the exact cause of the problem they just need to roll up their sleeves and commence resolving it. In some cases, they would be correct. However, in a lot of scenarios, there are things we need to do before making the repair to insure we are not causing any additional problems for our client or creating more work for ourselves later on.

The first thing we need to do is ask ourselves this. Is the solution we are about to implement the best way to resolve the problem? Sometimes simply replacing a part is all that is necessary for an effective repair. Other times it's the absolute worst thing you could possibly do. How so? To answer this question let's look back at our troubleshooting process.

During our learning phase we constructed a good, solid working knowledge of how the device operated. First, we identified the input, processing, and output functions of the device as well as the systems used to perform those functions. We know what the device uses as an input medium, the way in which that input medium is processed, and the desired output. We also observed the device during proper operation in its operating environment and made notes as to what the unit does under normal working conditions. Our observations informed us about a number of operational characteristics of the device including how it sounds, smells, looks, and feels.

During the identification phase, we used our working knowledge of the device and its operational characteristics to try and determine exactly what it was not doing. We asked the people who use the device a number of questions related to its normal operation as well as what it was doing at that time. We compared that information to our own observations of the unit in action in order to determine what was actually occurring.

Next, we searched for the root cause of the problem. We applied our working knowledge of the device and the identified problem and then, using the outside-in approach, and also taking into account any environmental factors that could play a contributing role, we were able to successfully narrow down the exact cause of the problem.

*4 Steps to Solving Your Problem*

After all that work it would truly be a shame if we were to just replace a broken part with a new one, or reconfigure its settings to how it was, and not take into account all of the information we obtained while investigating the problem. We must also consider the level and type of use—or abuse—this device receives in order to affect an appropriate resolution suitable for the operating environment in which it resides. For example, if this is a light or medium duty device operating in a heavy duty or industrial environment, simply replacing a broken part would be roughly the equivalent of putting a Band-Aid on a gaping wound. It may get the device back up and running, but you can be almost certain that you will be visiting this issue again in the not so distant future. Likewise, if this is a light duty device, an industrial-level fix would be quite unnecessary. A good rule to follow in this situation would be what I call the "finger rule." Very simply put, if I need two fingers to hold something together, three fingers would hold it together even better and with less effort than just using two. If I wanted to, I could use four or five fingers to do the same job, but at the end of the day, that would just be a waste of fingers. Just remember that the fix should fit the situation. To help insure a sufficient repair it's always okay to do a little more than what is needed but it's never okay to do less!

Many times in my career, while resolving a problem, I have also taken measures to improve upon the item in need of repair. Doing this not only fixes the current problem, but by improving on how the device was manufactured, installed, or configured, I lessen the chance that I will have to touch it again. This isn't always a necessity, but in cases where improvement is warranted, and you happen to be inherently lazy like me, you might as well do it now so you won't have to worry about it again! Plus, your client will greatly appreciate your going the extra mile to save them money on future repair bills. My point here is very simple. Make sure the resolution you put into place is at least sufficient for the operating environment or, better yet, more than what is required.

Why do I stress the importance of performing quality repairs? Well, if I had a dime for every time I have been called upon to fix something that had supposedly just been fixed by someone else, I would be able to give this book away for free! In these cases, 99% of the time the repair that was originally made was substandard for the environment and actual use of the device. Sometimes the fix was simply a part replacement. Other times it was a swap out for a new device of equal or lesser capabilities. However, most of the time it seemed the previous repair was made with utter incompetence. I've seen it all, too! I remember one time when I was called by a new client to look at a printer that was just repaired a few days prior by a supposedly reputable company. It was not advancing paper correctly, and after the other technician worked on it, the problem seemed to get worse. When I opened the printer up to look at the paper feed assembly I saw that it was being held together with a paper clip and surgical tape. I'm sure the look on my face when I saw that was priceless. It truly amazes me the lengths some so-called service technicians will go to create a substandard repair when doing things the right way can be done with just a little more effort.

## Unintended Consequences

The next thing you need to be made aware of is this lovely little premise I like to call the Theory of Unintended Consequences. In layman's terms this can be described as the simple act of fixing one problem and inadvertently creating another. How many times has this happened throughout my career, you ask? The truthful answer is that I can't count that high! Unfortunately for us, we can't always be sure that resolving one issue will not create another. The good news is that there are some steps we can take to help minimize the possibility of generating more work for ourselves than we originally bargained for. If we have followed the processes throughout this book, we have already gained most of the knowledge we need to help minimize these unforeseen mishaps later on.

The first step in your quest to avoid creating any unintended problems is to rely on your working knowledge of the device in question. I know, you are tired of hearing this and, to be perfectly honest, I'm getting pretty tired of typing it! Having said that, I cannot stress enough the importance of lesson number one in this book. Knowing all of the operational characteristics of a device can give you plenty of insight into the things you should focus on to effectively resolve an issue. It can also provide you with a huge list of items to be aware of that can adversely affect the unit's operation. I've described this concept as trying to fix an air leak in a tire by stapling a patch over the hole. Sure, the leak has been fixed, kind of, but I can definitely assure you that the problem will rear its ugly head again soon enough, and when it does, the problem will more than likely be much worse. Although you will never be able to totally eliminate the possibility that your solution for one problem will cause another, by relying on your working knowledge you should be able to identify most potential related issues before they happen.

**Inside-Out**

Another important lesson I've learned the hard way is this. Just because you found the root cause of the problem doesn't mean you found the actual root cause. In "Chapter 7 - Isolate the Problem," I discussed using the Outside-In approach to determine the root cause of a problem. While this method is the most effective way to narrow your search for the root cause of your issue, it can also be used in reverse. Why would we need to use this method in reverse? Because it can help us to determine if something within the device itself, or within the environment, could be contributing to the ultimate failure we have isolated. In Chapter 4 I briefly discussed examining the environment to determine if there were any factors contributing to the failure of our electric motor. I pointed out a number of things that should be examined to determine if the motor had any outside help in hastening its demise. This Inside-Out approach can be very useful in assisting you with determining if there are any seemingly unrelated issues that could have contributed to the actual root cause of the problem.

*4 Steps to Solving Your Problem*

Here is a great example of this method in action. I used to have a guitar amplifier that would blow capacitors seemingly for fun. A capacitor is an electronic device used to store electrical power. They can be found in just about anything that uses electricity in some shape or form and they are used for a number of different purposes. In the case of my guitar amplifier, this particular capacitor was being used to smooth out the flow of electricity supplied to the electronics after it was converted from alternating current (AC) to direct current (DC). The problem was that it would only provide this functionality for a few days at a time before literally exploding like a firecracker. I would replace it, it would work fine for a few days, and then, POP. I had replaced this capacitor so many times that I had turned the act of tearing the amplifier apart into an art form.

One day, while wrist-deep in amplifier guts, replacing this capacitor for what seemed to be the 1,000th time, I decided to snoop around a bit to see if I could find anything that might be helping this little bugger detonate. I followed the circuit paths from the component in question, making sure to test each component and its subsequent connections along the way. It didn't take long to find the culprit. There was a cracked solder joint on one of the terminals connecting the amplifier's power supply to the main circuit board. A quick test with my multimeter confirmed that the connection was good, but when I moved the circuit board, the connection would break. Guitar amplifiers tend to vibrate a little during normal use. If you are like me, and have a tendency to run your amp with as little resistance as possible on the Volume potentiometer (that means you play it really, really loud), that vibration is much more pronounced. The crack in the solder joint itself was extremely small, but when you add the vibration the amp creates while operating, it can cause that crack to open and close just enough that it produces a momentary short circuit. This condition is what caused the capacitor to fail. A quick zap with a soldering iron to heal the faulty connection ultimately fixed my exploding capacitor problem.

**Always Think, But Not Too Much**

The last thing we need to be sure to do will actually require the least amount of effort. We quite honestly do not want to overthink the repair. Sometimes the simplest solution is the best solution. You need to rely on your best judgement, as well as past experiences, to guide you on how to proceed with your solution. If the device you are working on simply needs a replacement part because the original part failed with no outside help, then replace it. Let's face it, sometimes things just break. Could be a manufacturing defect that got through quality control. Could've worn out over time. Or it just could be Monday (because we all know this type of thing happens on a Monday). However, if it needs a replacement part because the manner in which the device was installed is responsible for the part's failure, then make the improvement to the installation part of the repair.

And that's it! You did it! Congratulations!

# CHAPTER 9
# WAR STORIES

LET'S TAKE A look at some of the more interesting problems I have encountered in my career and how this method assisted me in resolving them.

For five years I was employed by a local county government IT department. I was assigned to the public safety complex and was responsible for computer systems and network connectivity used by the local detention center, sheriff's office, 911 center, emergency medical service, and emergency management service. My responsibilities did not end there. I also managed the county public safety radio system and radio and text paging systems, and I was the technology liaison for all local volunteer fire departments as well as local law enforcement agencies. I worked with other municipalities when needed.

When I started I knew there were a lot of new systems I had to learn so I jumped right in and began to investigate what I was dealing with. I concentrated on one system at a time and made sure I was comfortable with my knowledge of how it operated and how it was assembled, installed, and configured before I moved to the next thing. I also paid strict attention to how each system interacted with others within the environment. I made sure to document my findings as I went along. It took a while, but it wasn't too long before I had confidence that I could handle most problems as they occurred.

I was on call 24 hours a day, 365 days a year, so I knew I had to be able to take immediate action should the need arise and that I needed to be prepared at all times. I was issued a smartphone upon employment so I made sure that any vendor, service contract, or contact information I needed was with me at all times. I cannot count how many times this information came in handy!

## Cloudy Communications

One particular day, while out of town shopping with my then fiancé, I received a call from the 911 center that their internet access was down. While trying to walk the dispatcher through a few quick tests to determine where the issue was, the 911 lines lit up and he had to go and answer a call. Now most companies or agencies use internet service from local broadband providers such as cable or telephone companies. These vendors rely mostly on direct connect methods of connectivity for internet access. Not us. Our internet service was provided by a state agency via a wireless microwave link. For those who are not familiar, microwave networking is a form of wireless communication that is capable of very high speeds. Unlike radio communications microwave is a "line of sight"

*4 Steps to Solving Your Problem*

technology, meaning that each antenna on a microwave link must be visible to the other with no obstructions. These antennas are normally found on the lower sections of the towers because the lower part of a tower tends to flex or move less in windy conditions than the upper part, thus insuring a better and more stable connection.

I was able to have a coworker go to the 911 center and perform a few tests for me. While I was waiting for him to arrive on site I made a call to the state agency support line to have them test the microwave link to make sure it was up. Unfortunately, since this was a Saturday there was no one to take my call.

Now this was a Saturday in late November. It had been rather cold the prior few weeks but this particular day was bright and sunny and in the mid- to upper-60s. Our microwave link traveled from a tower behind our building to a tower in the next county over a rather wide body of water. The antennas were only about 120 feet off the ground.

So why were the weather conditions important enough to mention? Because at that particular time of the year the water temperature in the river drops, and since we'd had a two-week duration of rather cold temperatures it was a fair guess that this hastened the drop in water temperature. With the higher air temperature on this particular day my thought was that maybe there was a rather heavy fog bank over the river. I called a friend who lived near the water and asked him what the visibility was over the river and he told me that he couldn't see the river from his house as the fog had engulfed the end of his street. Why is this important? Because microwave signals and fog do not like each other! Normal, everyday fog has very little effect on microwave transmissions, but when the fog is abnormally thick, such as it was on that day, it can greatly increase the chance of connectivity problems.

When my coworker arrived on site I had him reset the connection equipment for the microwave link and when he did it worked great—for about six seconds—and then started to fail. This was a sure sign that the microwave signal had quickly deteriorated due to the heavy fog. I told him we were done working on it and then called the 911 center to let them know there was nothing I could do until the fog lifted, which was forecast to happen in a few hours. As predicted, when the fog subsided, the microwave system resumed communications and our internet access was restored.

From the time I received the call informing me that internet access was down to the time I isolated the problem was approximately 30 minutes. And I was able to do this from 60 miles away because I could call on my working knowledge of the system, I understood the environment in which the system was used, I determined what the exact problem was, and by eliminating possible causes, I determined the root cause. Unfortunately, I had to wait for Mother Nature to provide the fix for this one.

## Aaaaand We're Down

A good test for whether you have succeeded in the learning aspect of your troubleshooting process is when you are thrust head first into a critical scenario that no one, including yourself, has trained for. This particular instance forced me to concentrate not only on resolving the crisis but also on making sure appropriate measures were put in place that would minimize any potentially disastrous complications that could arise as a result.

It was the Saturday before Christmas and all through the town, not a radio transmission was heard because the system had gone down.

My wife and I were driving home from a family gathering, and when we were still roughly an hour from home, my phone began to ring. It was a call from the lead dispatcher at the 911 center mentioned earlier informing me that the entire county radio system (used by all first responders and city and county law enforcement as well as other agencies) was offline. The system that monitors the health of all the broadcast tower sites had reported that it was no longer in contact with the main site that controlled the system. This resulted in a reality that no one wants to be confronted with late at night, on a weekend, and before a major holiday. No one, be they law enforcement, firefighters, or emergency medical personnel, could use their radios for essential communications. Definitely not something you want to hear late in the evening when you are an hour away.

*War Stories*

My knowledge of the radio system was limited to my understanding that I didn't know enough about it to isolate the specific cause of an event. Instead I needed to gather as much information as I could to identify exactly what was happening and then try and eliminate as many possible causes as I could in order to assist the vendor's support team.

On this night task number one was to contact vendor support and initiate a support ticket. Next, I attempted to contact our local support technician as well as his service manager, but to no avail. I left messages for both. I then contacted the lead dispatcher at the 911 center to get the status of the situation only to find it had not changed. We discussed a plan of action which included informing all first responders and law enforcement of our plans for communications during the outage as well as recommending to all volunteer fire companies that they staff their stations with enough members to man their first-due units. Should there be a fire in their area it would be much easier to contact the station than it would each individual firefighter.

At this point I received a call from our service technician, who was also out of town. We began troubleshooting the issue with the information we had on hand, which was very little. We determined the problem to be at the main broadcast tower site and created a list of things for me to check upon my arrival. I called a friend who lived near the main tower site and asked him to check and see if the tower's marker lights were working. Once I confirmed that the lights were functioning I knew that we were not facing a power

*4 Steps to Solving Your Problem*

outage or generator failure and could rule that out as a possible cause.

About 20 minutes before I arrived at the site I contacted the 911 center again and requested a deputy to do a security check of the area. This tower site was out in the middle of a field and well out of town. A few months prior we had an incident where someone broke into the gated compound and proceeded to steal as much copper grounding wire as he could get his hands on. I wanted the deputy to make sure no one was attempting a repeat performance.

Upon arrival at the site I again confirmed that everything was powered up and running. Our service technician had connected remotely to the site but was not able to connect to either of the master system controllers. The radio system was fully redundant, which meant that there were actually two complete radio systems housed at the site. If the primary system experienced a failure of any kind a remote switchover module would immediately—and automatically—switch over to the secondary system. Once this happened, the system would send an alert message to the monitoring system informing us of the fault. Unfortunately for us, both systems had failed at precisely the same time. It appeared to us that the master controllers generated similar errors within seconds of each other; when the active controller failed the system switched over to the backup system, but it was already offline.

I restarted the backup controller and then tested the radio communications across the network. It worked. I

*War Stories*

restarted the active controller, switched over to the primary system, and tested it. It worked also. It was at that moment that the vendor's on-call technician arrived on site. Subsequent examination of log files confirmed what we had suspected. The log files revealed that the controllers both experienced a similar problem at precisely the same time, resulting in the failure of the entire system.

Okay, I'll admit that was a rather anticlimactic ending to what seemed to be a disastrous problem. As an IT consultant one of the most common remedies to everyday computer problems is to simply restart it. It is, however, somewhat astonishing that a simple reboot can also fix such a monstrous problem. What is even more difficult to believe is that such a major incident can be caused by the most minute of things (then again, all it takes is a little spark to ignite a forest fire). To be honest, neither the vendor nor myself could ever pinpoint an actual cause of the errors and, to my knowledge, the problem has never occurred since.

Now, remember the final lesson in Chapter 8? Always think, but not too much? This incident should serve as a perfect example of that. Yes, there was a system failure resulting in a catastrophic loss of service. Yes, there were errors documented on both controllers at precisely the same time. However, the errors did not indicate a possible cause of the problem within the system, only that a problem had occurred, and there was also no evidence of any physical damage or environmental issues that could be attributed as well. In this instance we simply chose to monitor the system for any further

*4 Steps to Solving Your Problem*

issues that could be related to this incident. Fortunately for us, there were none.

In this instance, my role was that of an intermediary. I used my rather limited working knowledge of the system in order to assist our vendor's service technician to identify the exact problem. Because I spent the time to learn as much as I could about the radio system I was able to provide our vendor the necessary information required to bring the system back to life. More importantly, having all of the necessary service information with me at all times allowed us to begin the troubleshooting process while I was still an hour away. This kept our total system downtime to less than 90 minutes.

These are extreme cases that most of you will probably never experience in your professional or daily lives but they should serve as examples of just what you can accomplish if you use this method.

**Sometimes You Have to "Vent"**

Not long ago a coworker dropped by my office to vent some frustration. She has a new (less than a year old) SUV and was frustrated because there was a problem with it the service technicians at the dealership couldn't figure out. Having dabbled in auto mechanics in the past I asked her to describe the problem to me.

She explained to me that sometimes, while driving the vehicle, she would get a strong odor of fuel. It didn't happen often but it happened enough that it was concerning to her. She said that it usually seemed to occur when she had to make a sudden stop or quickly slow down. She told me that the smell of fuel seemed to come right through the air vents. She had taken the vehicle back to the dealership numerous times but no problems were ever found. The computer diagnostics never reported any problems, no indicator lights on the dash ever lit up, and no one seemed to have a clue as to what was happening.

So I'm sitting there listening to all of this and the gears in my head start turning. The first thing that comes to my mind is a vent line might be cracked. Vents in a vehicle serve a multitude of purposes. In most cases they are there to relieve pressure that can build up within the engine and its supporting systems. There are also vents installed on the fuel tank to perform the same function. I started asking her some simple questions. "Have you noticed if your fuel mileage has dropped?" "Have you noticed the vehicle leaking anything?" And, "Does the smell actually come through the air vents?" Her response to the first two questions was no but she answered yes to the last question.

Receiving a "no" response to the first two questions answered one of the questions I had wondered to myself, which was why nothing showed up on the computer diagnostics performed by the dealership. Had there been a drop in performance with the engine the computer would have indicated this during diagnostics. I was pretty sure this venting system was not being monitored by the vehicle's computer. Knowing that all engines

*4 Steps to Solving Your Problem*

need to vent internal pressure in order to operate correctly, but also understanding that I have no clue how this is accomplished on modern engines (I used to work on cars when you could open the hood and actually SEE the engine), I told her to take the vehicle back to the dealer and have them look for a cracked vent line in the engine compartment.

My guess was the vent line that was damaged was either the crankcase ventilation line or perhaps a vent from the fuel evaporator. I was pretty sure the problem was the crankcase vent line because the fuel evaporator usually lives near the fuel tank, which is at the rear of most vehicles.

So, why would a cracked vent line in the engine compartment cause a fuel smell in the vehicle's passenger compartment? Well, my theory was that whenever she would have to slow down quickly or stop suddenly, any vapors that may have leaked into the engine compartment might find their way into the cabin fresh air intake. This fresh air intake, which is where the climate control system gets its fresh air from, is situated between the engine compartment and the passenger compartment in the general area of the windshield wipers on most vehicles. While the vehicle was moving any vapors or fumes from this cracked line would be directed out of the engine compartment far below and away from the fresh air intake. However, fuel vapors are lighter than air, and will travel upwards when not acted upon by another force, so these fumes could easily find their way to the intake if the car was to move slowly or come to a complete stop. This would explain how the smell would come through the vehicle's air vents.

A few days later she walked into my office and put a paper down in front of me. It was a Service Order from the dealership. She looked at me and said, "Read it." So I did. Turns out they found a crack in the crankcase vent line and replaced it. I shook my head and laughed as I gave her back the paper. Her next question to me is the reason why I am including this incident in my book. She said, "Now how is it a mechanic couldn't figure out this problem but an I.T. guy could?" Well, there is really a simple explanation for this.

Long ago, in the age before computers, people had to actually know how something worked in order to fix it. There were no built-in computers to operate, monitor, and diagnose problems, there were no 24-hour tech support lines, and there definitely were no internet resources like YouTube and Google. Believe it or not, there wasn't even public access to the internet. You had to not only have a strong working knowledge of something but you also had to know all the details of how something operated, all the systems involved in its operation, and how all of those systems interacted. Sound familiar?

Now I know what some of you might be saying. "I guess this guy just thinks he knows more than everyone else!" Well, no, I don't. I am confident that a trained automotive service technician knows a lot more about cars than I do. The reason I was able to figure out this problem isn't because I am super knowledgeable about

*4 Steps to Solving Your Problem*

cars. It is simply because I understand that computer diagnostics and error reports only tell part of the story (or in this case, none at all). After that, you are left with your working knowledge of the device and the information about the problem at hand. You see, these days we rely on computers for just about everything from basic operations to advanced diagnostics. The problems arise when people are trained to rely on computer diagnostics to resolve every problem. The mindset is that if the computer doesn't report a problem then there's a good chance there isn't one. This is apparently the obstacle my coworker was running into. The technician was trained to trust the diagnostics would give him all the information he needed to resolve the issue. In this particular case that didn't happen.

This brings me to another problem inherent with computers. Whenever a computer diagnostic reveals a problem or an error message pops up on your screen, do not immediately assume that the error being reported is an accurate representation of what is happening. As we learned earlier, the error message you are receiving may actually be a symptom of the real problem and not the problem itself. Ask any computer technician about the old Windows 98™ Memory Error and they will tell you that this error could indicate any one of a thousand problems, none of which had anything to do with a memory problem.

So how was I able to use our troubleshooting process to help my coworker resolve her problem? In this instance I used my knowledge of cars and how they operate, the various systems that all cars share as well as how those systems interact to take the information presented to me and identify exactly what was happening. Based on that information I was able to narrow down, or isolate, a possible cause. Knowing I was not the person who was going to fix this problem, I had the vehicle owner pass this information on to the service technicians at the dealership where they were able to find the problem and replace the necessary parts.

So, what is the moral of this particular story? Don't rely on computers to accurately tell you what is not working. LEARN!

# CHAPTER 10
# TIPS AND ADVICE

ALONG WITH THE methods described in this book there are some other things you can do to help you kick-start the process of acquiring a solid working knowledge of things. If you are like me you really like to dig into things and see how they work. And when I say dig in to something I mean just that. Tear it open and check out all the little gizmos inside. Now I'm not suggesting that you crack open your TV or dishwasher and start snooping around, but if you are ever in a situation where you have the chance to do just that, don't miss the opportunity! There is no such thing as too much knowledge and there is no substitute for actual hands-on experience. I have learned more about things just by tinkering around with them than any manual can possibly teach.

**Customers**

This was covered a bit in our chapter on people but it bears repeating. Quite possibly the best way to serve

*Tips And Advice*

your customers is to communicate with them in a manner they can relate to. It can sometimes be a daunting task to try and explain to someone with a limited knowledge of a particular thing exactly what is happening or how something works using plain English and not industry jargon. You must try and find a way to explain the situation to your customer so that they can gain an understanding of what is going on.

You may have noticed that this book is written in such a way that someone who is not familiar with a particular topic can understand the message I am trying to deliver without the use of a lot of language they may not understand. If you go back to the section on systems and look at the comparison I made in regard to how systems work together to perform a specific task, you will see that I used players on a football team to represent the individual systems. When describing the Input, Processing, and Output operations of a system I compared each system responsible for a certain task to an individual team member responsible for a similar task during the execution of a football play.

Using analogies like this to explain a situation to your customer goes a long way toward ensuring that they not only understand what is going on, but also trust that you know exactly what you are talking about. Let's face it, if the customer doesn't trust you, you can rest assured that they will not call you back for any future issues.

The most important thing to remember is to treat your customers in the same manner in which you expect to be treated! You can be the absolute master of time, space,

*4 Steps to Solving Your Problem*

and your profession, BUT, if you act like an arrogant jerk there is a good chance your first visit to a new client will also be your last. There are five basic building blocks for creating and maintaining good relationships with your clients. These simple, common-sense attributes that all of us are quite capable of displaying (even though a good number of us choose not to) can be easily remembered and referenced by employing the lovely acronym, PHART. Simply stated, you always need to be:

**P**olite

**H**onest

**A**ccommodating

**R**espectful

**T**imely

**Polite**

It's pretty self-explanatory, don't you think? Always be polite. Even if your customer is in a vile mood the simplest act of politeness can defuse any potential flare-ups between you. Remember, chances are they are not upset with you but with the situation. Don't take things personally. A simple "Yes sir/ma'am, no sir/ma'am" goes a long way.

*Tips And Advice*

**Honest**

Always be honest with your client. Period! If you don't know the answer to a question, tell them you don't know the answer but that you will get it for them as soon as you can. Whatever you do, never make up something just to be able to give them an answer. That is the best way to lose all credibility while looking foolish in the process. If, during a repair, something goes wrong and you require more time or supplies—or worse, something else breaks in the process of performing the repair—let them know. I would much rather have a customer be upset with me for accidentally breaking something and causing a delay in resolution than to have them get mad at me for not being honest about it. Everyone makes mistakes and no one has all the answers. Being honest about it builds trust and lays a solid foundation for a strong relationship with your client.

**Accommodating**

As a service provider, it is your job to solve as many problems for as many customers as you can as quickly as you can. Sometimes, however, there will be a job that needs to be completed for a client within a particular time frame, or there will be tasks you need to either perform or ensure they are performed, that are outside your normal scope of operations. In these cases, you need to understand that the client is not trying to be difficult or inconvenience you. Instead, their main focus is to minimize any impact the work you are about to

*4 Steps to Solving Your Problem*

perform for them may have on the operation of their business. The more accommodating you are to their needs, the more you will solidify your relationship with them.

**Respectful**

There is an old saying that goes "in order to be respected you need to be respectful." Unfortunately, a lot of people do not get a passing grade on this one. No matter how great you are at what you do, you will never be respected in your profession unless you are respectful of others—especially your clients! Always remember that while you are performing a task for your client, you are basically their employee. You need to understand that your client knows more about their business than you do, and it is your job to satisfy their requirements and not the other way around. There will be times when their requests may not make sense to you. Fulfill those requests anyway. It's okay to inquire as to why and to inform them of possible better and more efficient ways to accomplish the task, but if they want something done a certain way, be respectful of their decision and do it.

**Timely**

In business time is money and effective scheduling is a valuable key to success. As a service provider one of the most important things to remember is when your client

*Tips And Advice*

calls you, there is a good chance they need you as soon as possible. When you tell a client you will be there at a certain time, be there. If something comes up and you are going to be late make sure that you contact them immediately and inform them of any changes in scheduling. Above all, once you are on site, make sure you concentrate on the task at hand and resolve their issues in a timely manner. You are not there to chat with your client's employees about their weekend plans. Limit your phone use to what's necessary to complete the current task. If you need to take a business-related phone call that does not pertain to this particular client, be quick and to the point and return to the task at hand.

**Tools**

The second most important possession for a service provider, after knowledge, is the proper tools for the job. Make sure you have the appropriate tools of the trade in your possession at all times and make sure they are clean and in good working order. Nothing aggravates a client more than a service technician who cannot perform a task simply because he doesn't have the necessary tools. Even worse is a tech who has the tools but they are dirty, worn out, and unorganized. Failure to possess and properly maintain your tools is not only unprofessional in the eyes of your peers but to your clients as well.

In addition to possessing all of the necessary tools to perform your duties, you should be aware of two very useful tools available to anyone with an internet

*4 Steps to Solving Your Problem*

connection. Since almost all of us carry a smartphone these days, the two most important tools available at our fingertips any time we need them are, of course, Google and YouTube! I cannot tell you how many times I have been thrust into a situation I knew nothing about and within a very short period of time I was able to resolve the issue simply by searching these two sites. The amount of information available is staggering, to say the least, and is, without a doubt, an invaluable resource for you in a time of need.

YouTube is also an extremely useful marketing tool for your business. By using the technology you already have in your pocket (talking about your smartphone here), you can create short promotional and how-to videos and post them for the world to see. By linking these videos to your website and your social media accounts you are giving potential customers plenty of information about your business and the services you provide, and, above all, you are showing them that you have the knowledge to help them in a time of need.

Okay, let's get back to the subject of actual hand tools. No matter what industry you are in there are two very useful items you should have in your toolbox at all times. A multimeter! Now, I know what you are saying. "Didn't he just say there were TWO things I needed?" That's correct. You actually need TWO multimeters! What is a multimeter, you ask? It is a device that allows you to measure voltage, resistance (ohm), and amperage (current) on an electrical or electronic circuit. Why do you need two? Well, because there are two kinds of meters! Digital and analog.

*Tips And Advice*

A digital multimeter has an LCD numerical readout and is most useful when you need precise measurements. If you need to take any kind of electrical measurements and to have a precise reading down to the tenth or even hundredth of a volt, ohm, or amp then a digital multimeter is the tool for the job. An analog meter is simply not going to be able to accurately display that information.

An analog meter, one which uses a moving needle to display electrical measurements, is not only quite capable of giving fairly accurate readings but it can do something that most digital meters have a difficult time with. An analog meter is very useful in viewing fluctuations of voltage or current on a circuit. As voltage drops and rises on a circuit the needle will drop and rise accordingly, giving you a visual representation of what is occurring within the electrical circuit. Most digital meters cannot render an accurate representation of electrical fluctuations in this manner; therefore, it's best to have both at your disposal.

Now, I know some of you are saying that there are digital meters that can portray fluctuations in much the same manner as analog meters. You would be correct in saying that. The problem is that, in most cases anyway, that is a feature found only in the more expensive models of digital meters. Unless you have a need for such high-end equipment it is much more cost effective to purchase an inexpensive analog meter. Besides, it never hurts to have a backup just in case one stops working.

*4 Steps to Solving Your Problem*

**Step by Step**

When actively troubleshooting a problem, you must always remember to take things One. Step. At. A. Time. Unless your goal is to simply get this thing back up and running as quickly as humanly possible you should always limit your testing to a single component at a time. Whether testing or changing a physical component or adjusting a setting somewhere in the configuration, make sure you test the operation of the device after each individual change. If you do not, there is a good chance that you will replace a part or adjust a setting without knowing which change fixed the problem.

For example, let's say you are working on a computer and the problem is that it just shows a blank screen. If you immediately replace the monitor and the video card (or, in the case of most modern PCs, you replace the motherboard because the video is embedded within) and then everything works, you are left with the question of which component was actually causing the problem. You don't know, do you?

The proper way to determine what is causing a problem is by the process of eliminating one thing at a time. Step by step. In our scenario, you should have first tried the computer with a different monitor. If it still didn't work correctly you should have put the original monitor back on and then tried a different video card. If the new video card worked then the problem would be solved and, more importantly, you would know what the faulty component was. If not, then you would move on to the next thing. If you just jump in with both feet and start

*Tips And Advice*

changing a bunch of things at once without testing each change individually, and then it starts working, what fixed it? You may never really know unless you go back and undo everything you did one thing at a time.

You will be far more successful and much more efficient when you work towards a solution rather than backwards for an explanation. Working backwards to find what actually worked is never a good idea.

**Spotting Trends**

One thing I am always on the lookout for is trends. What is a trend, you ask? Well, in our case a trend is any time you have a particular problem that occurs on multiple devices in a specific area and time frame. Let's say you replace a power supply for a client's computer on Monday. On Wednesday, the same client has another computer with a bad power supply that needs to be replaced. Then, the following week, he loses two more. What we are now faced with is a trend of similar problems. Power supplies in a number of similar PCs are failing within a similar time frame and within close proximity. What could be causing this? Could it just be a coincidence? For the record, I don't believe in those! Chances are there is some underlying cause for this problem that needs to be addressed. Let's look a little deeper.

What we need to do is start looking for more similarities between the involved devices. Are they similar model computers? Were they purchased and installed at the

*4 Steps to Solving Your Problem*

same time or was there considerable time in between? Why does this information matter? Well, if they are similar machines, all purchased and installed within a short timeframe, then maybe these machines received faulty power supplies from the factory. A Google search on this model computer and potential power supply issues may reveal the answer you are looking for. Are other people with this particular model PC having the same issues? If you can't find an answer there then a call to the manufacturer's tech support line would be in order (notice I didn't say a quick call).

Now, let's say the machines in question vary in either model or brand and were purchased at different times. We are still having the same power supply issue, but since these power supplies do not reside in similar hardware we can rule out a possible manufacturer's defect. So what else could be the underlying cause? Well, if you have been paying attention to this book you know that there could be environmental factors at play. Perhaps there is an electrical problem within the building. Are all of the machines connected to the same circuit or different ones? Are the voltage and current at acceptable levels and stable or do they fluctuate? If so then the bad power supply problem is actually a symptom of something else that is failing within the environment.

Being able to recognize trends is a skill that not a lot of people possess but it is one that you can and should learn. You can build this skill by being observant of your surroundings and documenting everything either by making mental notes or physical ones. Understanding

*Tips And Advice*

that similar problems with different devices could be a sign of a larger issue is a valuable commodity and the type of skill that can increase your value as a service provider to both your employer and your client.

**Observe**

By far the greatest tip I can give anyone looking to improve their troubleshooting skills is to be observant. I've stressed this concept throughout this book and it is something that cannot be overstated. Be observant. Know your surroundings. Recognize any changes within the environment as well as any potential risk factor as it pertains to the equipment you service and to the safety and productivity of your customers.

And always be on the lookout for trends!

If you are troubleshooting a piece of software and browsing through menus, configuration files, and settings windows, always make a mental note of what you see. Even if the information you are looking at has no bearing on what you are currently working on there is always a chance that it may come in handy later.

**Documentation**

This section consists of two parts. First let's concentrate on product documentation. This consists of the manuals, product warnings, warranty cards, and all the other seemingly endless papers that come with just about every product on the market. Always make sure you know where the product documentation is for whatever you are working on, and that you have access to it. If you are the person who installed the equipment make sure that the documentation is organized properly and in a safe place for easy access should the need arise, and make sure the client understands the importance of properly maintaining it. Far too many times I have needed some information from these innocuous little pages only to not be able to find them, thus doubling or tripling the time needed to resolve an issue.

Now let's have a look at the other type of documentation. This is the documentation you create to assist you with maintaining your client's equipment. Make sure you keep some sort of documentation on each of your clients to track problems, configurations, notes on environmental and infrastructure issues, and any other pertinent information. Properly documenting each visit to a client may take a bit of time at that moment but it can save you a lot of time later when problems arise. Keeping accurate notes on previous problems can help you resolve similar issues that you may be facing currently, saving you time and your client money. It is also essential to assist you in spotting trends that may be occurring that, ordinarily, you might have overlooked.

## Working Well With Others

There will be times when you will be working with other vendors or contractors on the same project. Always, and I mean ALWAYS, introduce yourself and

*Tips And Advice*

maintain contact with all of the other contractors on the project and do so in a professional manner. When working with other contractors proper and open communication is the key to a successful conclusion. There will be times when you will need to work around the schedules of others and there will be times when they will have to work around yours. Always be courteous, understanding, and willing to go the extra mile to work with them. There is a good chance they will return the favor and be even more willing to work with you should the opportunity arise again. I've even had other contractors that I have worked with on past projects recommend me to prospective clients for new ones.

**Rules, Not Laws**

Someone once told me that rules were made to be broken but laws were not. I'm not sure that is true, but, be that as it may, it leads me to my next tip. And that is, even though the methods I have outlined in this book work for me, there is no law that states you have to follow this procedure verbatim. In fact, if you go back and reread Chapter 1 you will see where I said that I have used this method to resolve "just about every" problem I have encountered. The truth is there is no one way to troubleshoot or resolve problems and you may find that there are parts of my method that work for you when combined with parts of other methods. The trick is, when entering into a situation where you are the one charged with resolving a problem, always keep an open mind about what you see. Just because you ran into a similar incident a few days ago doesn't mean this incident is indeed the same. I have seen too many times when someone entered into a situation with the mindset that the problem he was facing was due to X, Y, and Z based solely on what he had experienced in the past and not what he was experiencing then. Experience is a wonderful thing but without investigating, identifying, and isolating the actual cause of the current crisis, you could cost the customer an enormous amount of money and time. Don't be that guy!

**You Can't Win Them All**

And last, but certainly not least, I leave you with this. Once in a while, no matter what you do, no matter how much you investigate, test, update, or swap out, you will experience a problem you simply cannot figure out. No matter how much you concentrate on it, how long you work on it, or how many choruses of the four-letter serenade you sing to it, you will not be able to resolve it. When that happens, please do not worry. You are not alone! As talented as I think I am as a service provider and a troubleshooter, I have come across a few things in my career that have left me dumbfounded, and to this day, make me scratch my head in utter disgust. It happens to us all.

Want an example? One time I had a client who could not get a computer connected to the network. Sounds simple, right? Well, it was anything but. I checked the network adapter on the computer, updated the software for it, cleared the TCP stacks and reset the Winsock in

*Tips And Advice*

Windows… nothing! I then moved that computer to another location and it connected without issue. Aha!!! The problem was in the cabling. I tested the patch cable. Good. I tested the cable from the network jack to the switch in the data center. Good. Then, I certified both cables at 1 gigabit speed. All was good. I checked the connectors and jacks on the cables and replaced them all. All tests were fine. I put the computer back at its original location and plugged it in. Nothing! I put another computer in that location. Once again, goose egg (nothing!). I then tested the cable yet again to make sure all eight wires were connected with no shorts or breaks. It tested successfully. I traced the cable through the ceiling and back to the data center and checked it for kinks or damage. No problems. This went on for two hours. Finally, I gave up and ran a new cable through the ceiling back to the data center. Then, and only then, did the computer connect successfully to the network.

It happens. It's frustrating when it does, but the last thing you should do is to let it get you down or start questioning your abilities. Some people you just can't reach and some things you just can't figure out. That's the way it is. When it happens simply cut your losses, call in the cavalry, and keep on truckin'!

# CHAPTER 11
# IN CONCLUSION

IT'S 35 DEGREES outside and raining, and you are 20 minutes away from somewhere you needed to be 10 minutes ago. You get in your car, put the key in the ignition, turn it, and nothing happens. So, what do you do now?

Well, if you had paid attention to the information provided within the hallowed pages of this book you would know exactly what to do next! You would already have gained a working knowledge of your car, understood the car's Input, Processing, and Output processes, and had a basic understanding of the systems involved with those processes. You would then use that information to assist you in identifying exactly what the car is not doing (or at least be able to narrow it down to a few possibilities). Then, after knowing what the car is not doing correctly, you would be able to isolate the exact cause of the problem and take the necessary steps to conquer it.

Throughout this book it has been stressed, time and time again, that the most important thing in your career is a solid working knowledge. In fact, knowledge—of any

*In Conclusion*

kind—is the key to success in your career as well as in life itself. Without a solid foundation to rest upon a building will most likely crumble. Likewise, without a solid foundation of knowledge your life and career will likely do the same. Anyone in the business of solving problems for others understands the fact that you can never know too much. Knowledge is the difference between being merely satisfactory at your job and being highly successful and sought after. Above all, in order to be successful at anything, you must remember that learning is the one thing you can never stop doing! Make an effort to learn something new each and every day. It doesn't even matter if it's useful knowledge. Even "useless" little tidbits of information can come in handy once in a while. As long as you continue to learn new things on a daily basis you will achieve success no matter what you choose to do.

As a service provider, it is your job to know more about the things you support than anyone else, including the end user and the person who sold the product to them. Think about that for a second. Let that sink in nice and deep. It makes a lot of sense when you stop and think about it. I mean, what does a salesperson actually know about this product? They know enough to be able to sell it to a customer. (Sometimes they don't even know that much. I can't count how many times I've had a salesperson tell me he needed to check with his "engineers" in order to answer what I considered to be a simple question.) A salesperson's knowledge of a product is usually limited to its superficial aspects. They may have used or operated the device before and know what it is supposed to do and how it can benefit their

*4 Steps to Solving Your Problem*

client. As a service provider, it is your job to not only know the superficial aspects of a device but also its inner workings, how it needs to be installed and configured, its appropriate operating environment, and how to fix it when something goes wrong.

Another thing we need to be mindful of is that, no matter our level of knowledge about something, we should not let it deter us from learning even more. Never feel like you know everything because, I can assure you, you don't. There will always be someone out there who knows more about something than you do. You should make it a point not to compete with them but rather to learn from them. Who knows, maybe you can also teach them a thing or two. Let's leave the competition to the athletes and the salespeople, shall we?

What if we are presented with a situation we are not familiar with? What if we have been asked to try and resolve something we have no experience with? One of the things we learned is that our working knowledge of something can also be used to help us figure out how to resolve issues with things we have no knowledge of. In a case like this we need to analyze what this particular device does, determine its input, examine how it processes that input and what the desired output is, and then compare this functionality to something we already have knowledge of. It's important to understand that even though two different devices may be used to accomplish two completely different tasks, they often share the same principles and methods. The differences between the two devices are in the way those methods and principles are applied. Remember, a hair dryer and a

*In Conclusion*

clothes dryer are essentially the same tool designed to accomplish different tasks.

And then there are those we must deal with on a daily basis during our quest to rid the world of its problems. We must understand how to deal with and effectively communicate with those whom we serve, with their many personalities and quirks-a-plenty. You will learn quickly that getting information out of some of your clientele can be more difficult than resolving the problem itself.

Here's a funny story about people. I once had a coworker who eventually went on to become a pastor (this incident probably cemented his decision to do so). A few days into his tenure we were having a discussion about the clients on his schedule for that day. I jokingly made a comment in reference to one client about believing he was proof that God had a sick sense of humor. He asked me why and so I described this particular client he was to visit. He was an interesting individual to say the least. He was mildly entertaining and just chock full of ridiculously inappropriate comments, none of which ever seemed to relate to another. Upon returning from his appointment with this client, he popped his head into my office and somewhat somberly stated, "I tend to agree with your assertion about God's sense of humor!" He then shook and lowered his head and quietly walked away. He wouldn't elaborate on his experience with said client. He did, however, ask that I not send him there again. My imagination is pretty vivid but I'm sure I can't conjure up anything close to what he experienced that day.

*4 Steps to Solving Your Problem*

All the people skills in the world will not prepare you for having to deal with customers who are in the midst of some sort of crisis. As frustrating as it can sometimes be you should always keep a level head on your shoulders and do the best you can to divert your customer's focus to the actual problem at hand and not the chaos that they, or someone else, might be creating because of it. We know that most of the problems we encounter can be dealt with fairly quickly but most customers do not know that. You will be amazed at how a simple phrase like "this doesn't look too bad" or "I've seen much worse than this" can quickly defuse a tense and sometimes chaotic scene.

Assuming you have been able to determine from your client what the actual problem is, you then need to use your knowledge of how something works to help you determine what exactly it is not doing. You are not trying to determine what is wrong with it; instead, you are trying to determine what function it is not performing. Always avoid asking someone what is wrong. Not only will the response not be an accurate representation of the problem but you may be told a little more about the device than you care to hear. Instead, ask them something a little more direct like, "Can you tell me what it is not doing?" The response you will receive from this question will provide you with an answer that is much more useful.

Be thorough and vigilant when attempting to identify a problem, because the issue you find may very well be just a symptom of another problem altogether. Once you identify what the device is not doing you must take the

*In Conclusion*

time to examine other factors, such as the surrounding environment, to ensure that you are, indeed, seeing the root of the issue.

Upon identifying the problem we must then set our sights on determining its root cause. It is at this point that we will use an Outside-In approach to eliminate possible causes beginning with the most general aspects of operation. As we rule out potential causes we narrow our focus to more and more specific possibilities until we find the culprit. Again, once we find the root cause of the problem we will further examine other factors that may have contributed to the failure. As we noted before, what we have determined to be the root cause of the problem could actually be a symptom of something else entirely.

The final step in our process is to conquer the problem. Care must be taken to effect a proper resolution as there are a great number of factors you need to consider. You learned that sometimes just replacing a part or changing a configuration can be the worst thing you can do. Never take corrective measures until you have taken the time to examine your options based on the working environment in which the device resides. If you notice that a particular part has failed due to normal wear simply replacing that part should be all that is necessary. However, if a part has failed prematurely there is a good chance that your replacement should be of a better grade or quality than the original. If the failure was caused by environmental factors then measures to neutralize those factors must be considered as a part of the overall resolution.

*4 Steps to Solving Your Problem*

If your resolution involves making any improvements to the device you must make sure that any changes you make do not inadvertently create new and unforeseen issues. It cannot be stressed enough that any changes made to a device must be analyzed thoroughly before implementation. Although you can never completely rule out the possibility that a modification will not cause any unintended consequences, properly vetting proposed changes can help minimize those chances.

In closing, allow me to say that I am deeply honored that you have taken the time to purchase and read this book! I can assure you that I have no plans for it to be my last. It is my hope that this will be the first in a series of books tailored to the customer service industry. The focus of this series will be on clear, common-sense solutions designed to assist corporations, small businesses, and individuals alike in their journey to be the best customer service providers they can be!

Throughout my life I have always done my best to help others in need. I sincerely hope this book further accomplishes that quest. Whether you realize it or not, those who embark on a career path that focuses on helping others are a special breed of human being. By choosing this career you are essentially devoting your life to helping others. And trust me when I tell you this, there is no better feeling in the world than having the ability to help someone when they can't help themselves!

*In Conclusion*

Even though I've spent this entire book dispensing advice of a somewhat technical nature, there is a bit of personal advice I would like to pass on. That advice is to aim for the Heavens but be sure to keep your head out of the clouds! What does that mean, you ask? It means never settle for anything less than being the best you can be, and never lose sight of where you came from, who you are, and, more importantly, where you are going.

Contrary to popular belief, life is not a competition. You should never strive to be as good as or better than someone else, but instead, you should always strive to be the very best you can be. Using someone else's assumed success as a guide to what your goals should be or what you should aspire to actually sets unnecessary limitations on you! The last thing anyone should ever do in life is limit their opportunities based on the achievements, be they actual or implied, of others. Success isn't about how much money you make, how famous you are, or how many things you have. The true meaning of success is being able to look at yourself in the mirror at the end of every day and being happy with the person staring back at you.

Until we meet again…

www.ingramcontent.com/pod-product-compliance
Lightning Source LLC
Chambersburg PA
CBHW070625300426
44113CB00010B/1659